Dynamic Binary Modification

Tools, Techniques, and Applications

Synthesis Lectures on Computer Architecture

Editor
Mark D. Hill, *University of Wisconsin*

Synthesis Lectures on Computer Architecture publishes 50- to 100-page publications on topics pertaining to the science and art of designing, analyzing, selecting and interconnecting hardwarecomponents to create computers that meet functional, performance and cost goals. The scope will largely follow the purview of premier computer architecture conferences, such as ISCA, HPCA, MICRO, and ASPLOS.

Dynamic Binary Modification: Tools, Techniques, and Applications

Kim Hazelwood

ISBN: 978-3-031-00604-3   paperback
ISBN: 978-3-031-01732-2   ebook

DOI 10.1007/978-3-031-01732-2

A *Publication in the Springer series*
SYNTHESIS LECTURES ON ADVANCES IN AUTOMOTIVE TECHNOLOGY

Lecture #15
Series Editor: Mark D. Hill, *University of Wisconsin*
Series ISSN
Synthesis Lectures on Computer Architecture
Print 1932-3235   Electronic 1932-3243

Dynamic Binary Modification

Tools, Techniques, and Applications

Kim Hazelwood
University of Virginia

SYNTHESIS LECTURES ON COMPUTER ARCHITECTURE #15

ABSTRACT

Dynamic binary modification tools form a software layer between a running application and the underlying operating system, providing the powerful opportunity to inspect and potentially modify every user-level guest application instruction that executes. Toolkits built upon this technology have enabled computer architects to build powerful simulators and emulators for design-space exploration, compiler writers to analyze and debug the code generated by their compilers, software developers to fully explore the features, bottlenecks, and performance of their software, and even end-users to extend the functionality of proprietary software running on their computers.

Several dynamic binary modification systems are freely available today that place this power into the hands of the end user. While these systems are quite complex internally, they mask that complexity with an easy-to-learn API that allows a typical user to ramp up fairly quickly and build any of a number of powerful tools. Meanwhile, these tools are robust enough to form the foundation for software products in use today.

This book serves as a primer for researchers interested in dynamic binary modification systems, their internal design structure, and the wide range of tools that can be built leveraging these systems. The hands-on examples presented throughout form a solid foundation for designing and constructing more complex tools, with an appreciation for the techniques necessary to make those tools robust and efficient. Meanwhile, the reader will get an appreciation for the internal design of the engines themselves.

KEYWORDS

dynamic binary modification, instrumentation, runtime optimization, binary translation, profiling, debugging, simulation, security, user-level analysis

To my husband Matthew
and our daughters Anastasia and Adrianna
for their patience and encouragement
while I worked on this project,
and for their ongoing love and support.

Contents

Acknowledgments

Much of the content of this book would not be possible without some of the great innovators in this field. Some people who stand out in my memory include Robert Cohn, Robert Muth, Derek Bruening, Vas Bala, Evelyn Duesterwald, Mike Smith, Jim Smith, Vijay Janapa Reddi, Nick Nethercote, Julian Seward, Wei Hsu, CK Luk, Greg Lueck, Artur Klauser, and Geoff Lowney. I should also thank the countless contributors to Pin, as well as the contributors to the many projects that preceded and formed the foundation for those projects listed throughout this book.

I would like to thank Mark Hill for approaching me and encouraging me to write this book, as well as the feedback and support he has provided throughout the project. Additionally, I would like to thank Michael Morgan for providing me the opportunity to contribute to this lecture series and for doing his best to keep me on schedule.

I should also acknowledge the people who advised me against writing this book prior to tenure (who shall remain nameless). Although I ultimately ignored that advice, I do know that it was well intended, and I always appreciate those who take the time and express enough interest to offer advice to others. Meanwhile, I always tend to fall back on the mantra to "Keep asking for advice until you get the advice you want."

Kim Hazelwood
March 2011

CHAPTER 1

Dynamic Binary Modification: Overview

Software developers, system designers, and end users all may have important reasons to observe and potentially modify the runtime behavior of application software. Dynamic binary modification systems provide this functionality, while hiding the complex engineering necessary to make the entire process work. These systems provide access to every executed user-level instruction (including calls through shared libraries, dynamically-generated code, and system call instructions themselves) while providing the illusion that the application is running natively. Aside from any potentially observable runtime or memory overhead, the application appears to behave identically to a native execution, including instruction addresses, data addresses, and stack contents. Some of the more commonly known dynamic binary modification tools today include Valgrind, Pin, and DynamoRIO.

From a high level, providing this high degree of control over a running guest application requires interjecting a software layer between the application and the underlying operating system at the process level. Contrast this notion with that of *system virtualization tools*, such as VMware or VirtualPC, where an additional software layer lies beneath an operating system and virtualizes all running applications including the operating system itself. By instead focusing on a single application, providing the functionality at a per-process level allows for individual inspection or customization of the application of interest, while allowing the overhead to be more easily amortized upon reaching an application steady state.

Dynamic binary modification systems operate directly on program binaries with no need to recompile, relink, or even to access the source code of the guest application. This is important in many cases because it allows a user to analyze and manipulate legacy code, proprietary code, streaming code, and large code bases in a straightforward and robust manner.

A dynamic binary system operates as the guest application executes (dynamically at runtime), performing the required program modification on demand. Contrast this notion with a static approach where an application binary would be regenerated in its entirety either at the start of execution or sometime prior to execution. Operating dynamically allows the system to be much more robust for a variety of reasons. Unlike static approaches, a dynamic system can handle applications that generate code dynamically, applications that contain self-modifying code, and it allows access and the potential to modify all shared libraries that an application may call. Furthermore, dynamic approaches can handle the intricacies of an architecture like the x86, which has variable-sized instructions combined with mixed code and data, making the task of static instruction decoding difficult if not impossible,

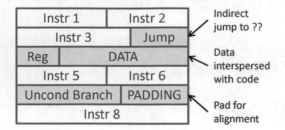

(a) Dynamic x86 Decoding Example   **(b) Static x86 Decoding Example**

Figure 1.1: The code-discovery problem that arises from variable instruction lengths, mixed code and data, and indirect branches (Smith and Nair [2005]). On the left, we see the correct interpretation of the binary as determined by dynamic analysis. On the right, we see an incorrect interpretation as determined by static analysis.

as shown in Figure 1.1. Finally, by operating at runtime, the dynamic binary modifier only targets the portions of the guest application code and program paths that actually execute.

1.1   UTILITY

Dynamic binary modifiers have been used for a wide variety of reasons, many of which the designers of the original systems had never envisioned. Users now span the subfields of computer architecture, compilers, program analysis, software engineering, and computer security. We'll take a high-level look at some of these motivating applications in the following sections, and we will follow up with detailed examples in Chapters 3–5.

Utility For Application Developers   Software engineers have a myriad of reasons to require a detailed understanding of the software systems they develop. While this performance analysis can be done in an ad-hoc manner, dynamic binary modification enables a more systematic approach to software profiling. Rather than mining massive amounts of source code, potentially missing key instances, developers may instead analyze the runtime behavior of their applications using a simple API and minimal profiling code. For instance, they can analyze all of the branches in their program (and all shared libraries it calls) using one or two API calls, or they can classify all of the instructions executed using a small number of calls.

Developers may also wish to perform systematic debugging of their software. For instance, they may wish to ensure that every dynamic memory allocation has a corresponding deallocation. Using binary modification to dynamically record every allocation, this goal can be achieved with very little developer effort.

Utility For Hardware Designers An interesting application of dynamic binary modification is emulating new instructions. Given that the binary modifier has access to every instruction before it executes, it can recognize a new instruction that is currently unsupported by the hardware. Instead of executing that instruction and causing an illegal instruction exception, the system can emulate the new behavior while measuring the frequency of use of the new instruction. In fact, a similar approach can be used to mask faulty implementations of machine instructions, by dynamically replacing those instructions with a correct emulation of that instruction's desired behavior.

A more general application of dynamic binary modification is to generate live traces for driving simple simulators. For instance, a user can write a simple cache simulator by instrumenting all memory accesses in a guest application. Memory access data can either be written to a file to drive an offline simulator, or it can be piped directly to a running cache simulator. Similarly, a branch prediction simulator can be written by instrumenting all branch instructions to record the source address, target address, and branch outcome. Finally, a full-blown timing simulator can be written by instrumenting all instructions to record any information necessary for driving a timing simulator, though it is only possible to measure the overhead of committed instructions using this mechanism. Committed instructions are all that are visible to a software-level binary modification tool.

Utility For System Software Designers Yet another application of dynamic binary modification is the ability to add and enforce new security or privacy policies to existing applications. A user may wish to enforce that applications do not overwrite instructions or jump to locations that have been classified as data. The ability to observe and potentially modify every application instruction prior to executing that instruction makes these tasks straightforward.

The motivating applications listed in this chapter attempt to demonstrate the wide variety of possibilities that arise when a user is given the ability to observe or modify every executing instruction. Each of these examples is described in deeper detail, with sample implementations and output, in the later chapters. Meanwhile, all examples presented in this book simply serve to scratch the surface of the potential of this technology.

1.2 FUNCTIONALITY

A user-level dynamic binary modification system is generally invoked in one of two ways. First, a user may execute an entire application, start to finish, under the control of the system. This approach is well suited for full system simulation, emulation, debugging tools, or security applications where full control and complete code coverage are paramount. In the second invocation method, a user may wish to attach a binary modification engine to an already running application, much in the same way that a debugger can be attached to/detached from a running program. This method may work well for profiling and locating bottlenecks, or simply to figure out what a program is doing at a given instant.

Whatever the invocation method, most binary modifiers have three modes of execution: interpretation-mode, probe-mode, and JIT-mode execution. In an interpretation-mode execution, the original binary is viewed as data, and each instruction is used as a lookup into a table of alternative instructions that provide the corresponding functionality desired by the user. In a probe-mode execution, the original binary is modified in-place by overwriting instructions with new instructions or branches to new routines. While this mode results in lower runtime overhead, it is quite limited, particularly on architectures such as x86, and therefore JIT-mode execution ends up being the more common implementation. In a JIT-mode execution, the original binary is never modified or even executed. Instead, the original binary is viewed as data, and a modified copy of the executed portions of binary are regenerated in a new area of memory. These modified copies are then executed in lieu of the original application. Both probe-mode and JIT-mode execution models are discussed in more detail in Chapter 2, while their internal implementation is discussed in Chapter 6. Interpretation is not discussed as the overhead of interpretation prevents it from being widely used in these systems.

Once the user of a dynamic binary modification tool has control over the execution of a guest application, they then have the ability to incorporate programmable instrumentation into that guest application. They can define the conditions under which to modify the application (e.g., upon all taken branches) as well as the changes they wish to make (e.g., increment a counter or record the target address). From there, the binary modifier will transparently inject the new code into the running application, taking care to perform supporting tasks, such as freeing any registers necessary to perform the function, but otherwise maintain the system state that the application expects. The level of transparency may vary by system – e.g., some systems will avoid writing to the application stack while others will borrow the application's stack temporarily. Either way, most systems do ensure that the observed state is as close as possible to that of a native run of the guest application.

1.3   SYSTEM PERFORMANCE

One of the most common questions that arises with respect to the notion of dynamic binary modification is the performance overhead of modifying running applications. Unfortunately, there is no simple answer to this question. The overhead observed by a user is highly dependent on a large number of factors, including the features of the guest application being observed and the particular modifications that the user wishes to implement (Uh et al. [2006]).

First, let's focus on the internal system overhead for acquiring and maintaining control of the guest application. Unlike a standard interpreter, the internal engine of a dynamic binary modification system will use a variety of techniques to leverage previous work. For instance, if a user wishes to modify a loop, the system will modify the loop once at the start of the program, and then it will execute the modified loop from that point forward. Therefore, nearly all of the system overhead can be amortized over a long-running stable application that has a small code base (i.e., primarily, loop-based codes). The worst relative performance would come from an application that runs for only a few seconds or less since there would insufficient time to amortize any start-up costs of modifying the code. Another challenging guest application would be one that contains a large number of features that are expensive to manage, such as indirect branches or self-modifying code. For these reasons, the baseline overhead (with no actual changes made to the application) can vary from nearly zero for some SPEC benchmarks, up to several orders of magnitude for some synthetic benchmarks. If we focus on the SPEC2006 reference inputs in isolation, overheads tend to hover around 30% with one order of magnitude range in each direction. Chapter 6 provides a look into the internal workings of the system to provide more insights into the specific sources of system overhead.

However, for most uses of dynamic binary modification, the majority of the overhead comes from the modifications requested by the end user. If a user wishes to perform invasive, detailed analysis of each and every instruction that executes, the overhead of performing this task will be quite high, and it would be mostly dependent on their particular implementation of their desired feature. Therefore, helping a user understand how to optimize their implementation is often much more helpful than helping them understand the internal workings of the system itself. As a result, Chapter 2 highlights the common pitfalls that an untrained user may inadvertently experience and provides the background information necessary for solving those problems.

1.4 HIGH-LEVEL SUMMARY

Dynamic binary modification systems provide a conceptually simple but powerful platform for building a wide variety of useful tools. These systems are in use today by numerous researchers to understand or change program or system behaviors. Operating at a user level, these software systems focus on a single guest application at a time and perform all essential tasks at runtime as the guest application executes. Effectively using these systems requires an understanding of the various execution modes available, and the impact of each choice in terms of functional limitations and resulting performance.

CHAPTER 2

Using a Dynamic Binary Modifier

This chapter is intended to provide the basic knowledge and terminology necessary to get started using a dynamic binary modification system. The content is geared toward potential users of these systems, be they computer architects, software engineers, or systems researchers. Advanced details about the internal implementation of these systems will not be covered until Chapter 6, and it will be geared toward those wishing to do research or development on binary modification systems directly. Instead, the discussion in this chapter will remain high level and system independent whenever possible. Specific systems, such as Pin, DynamoRIO, or Valgrind, will be discussed as necessary, always from a system usability perspective.

2.1 HEAVYWEIGHT VS. LIGHTWEIGHT CONTROL

One of the first decisions that a user of a binary modifier will need to make is the amount of code coverage required for their intended application and the amount of performance overhead they can tolerate. These decisions will lead them to the choice between modification-in-place (probe-mode execution) or modified-copy (just-in-time code regeneration), which are two modes of execution available in most systems. This choice has many implications both on the performance and the functionality fronts.

2.1.1 JIT-MODE EXECUTION

The most common execution mode (and the default on most systems) is to use a just-in-time compiler to regenerate a modified copy of a small chunk of instructions, immediately prior to executing those instructions. The modified instructions are then cached in a memory-based *software code cache*[1] from where they can be reused for the remainder of the execution time. Therefore, in the ideal case of an unlimited cache, each static instance of an instruction is only modified once, but the resulting modified code is leveraged from that point forward.

JIT-mode execution is the most robust execution model, and it performs best on guest applications that experience a great deal of code reuse (loop-based codes) as the overhead of regenerating the cached copy can then be amortized throughout the execution time of the program. For very

---

[1]The software code cache is also called a *translation cache* in some sources.

short-running programs and/or programs with few iterations, it becomes difficult to amortize the overhead of just-in-time code regeneration.

2.1.2   PROBE-MODE EXECUTION

One of the lesser known execution modes is that in which the original binary is patched in memory and, therefore, the modified version is used for the duration of that execution time, rather than a cached copy. The overhead of this technique is much lower since the bulk of execution time is spent executing native code. However, there are a number of restrictions to using this mode, particularly on the x86 architecture, and therefore many dynamic binary modifiers do not support this mode at all. The main restriction is a result of the variable-instruction-length nature of the x86 architecture. Adding new functionality generally involves overwriting an existing instruction with a branch instruction that jumps to a new routine. However, not all instructions can be overwritten by a branch, because the x86 branch instruction is at least 5-bytes long (longer if the branch target is beyond a certain distance away). Therefore, all instructions that are shorter than 5-bytes cannot be modified without also affecting part of the subsequent instruction. Overwriting two instructions at once is unsafe for two reasons. If any other branch targets that second instruction, it would now jump into the middle of a branch instruction, and it could crash. Second, for multithreaded applications, the system cannot guarantee that another thread is not stalled between the two instructions, and it would resume in the middle of the newly modified instruction. To circumvent these architectural features, the systems that do support probe-mode execution only allow the application to be modified at a small set of locations, namely function call boundaries and user-level code. These restrictions do not exist for architectures with fixed-length instructions, such as ARM, MIPS, or Alpha, as any arbitrary instruction can cleanly be overwritten by a branch to an instrumentation routine.

2.1.3   PERSISTENT BINARY MODIFICATION

Regardless of the execution mode employed, the code modifications that occur during one execution of a program are discarded and are regenerated during subsequent runs. Researchers have explored the notion of *code persistence* – saving the modified code generated during one execution for use in subsequent executions (Bruening and Kiriansky [2008], Hazelwood and Smith [2003], Reddi et al. [2007]). However, in most cases, the general process was deemed too fragile and complicated for widespread implementation and use, and it is therefore not supported on most systems. More details on persistent code modifications are available in Chapter 6.

2.2   LAUNCHING THE SYSTEM

There are two fundamental methods for taking control of a guest application. First, the user can specify that the entire program will run under the control of the dynamic binary modification system, from the first to the last instruction. And second, the user can attach the system to an already running application and later detach from that application, allowing it to continue to run natively. The specific execution mode that the user chooses will depend on the functionality they wish to implement, and whether that functionality is conducive to relinquishing control of the guest application.

2.2.1 COMPLETE PROCESS CONTROL

The standard way to instrument or modify a program is to run the entire program under the control of the dynamic binary modifier. The system will take control before the first guest application executes and will maintain control until the guest application terminates. From a user standpoint, launching a dynamic binary modifier is very straightforward. Below are the necessary commands for launching Pin, DynamoRIO, and Valgrind on Linux.

```
//Launching Pin on Linux (JIT mode)
pin <pinargs> -t <pintool>.so <pintoolargs> -- <app> <appargs>

//Launching DynamoRIO on Linux (JIT mode)
drrun -client <client>.so 0 "" <app> <appargs>

//Launching Valgrind on Linux (JIT mode)
valgrind <valgrind_args> --tool=<toolname> <app> <appargs>
```

The commands demonstrate that there are three facets to the binary modification system: (1) the system engine itself, (2) a user-specified plug-in tool, and (3) the guest application and its arguments. Furthermore, other than prepending its command with two additional commands, the guest application is launched the same way it would have launched natively, including all paths, flags, and arguments.

On Windows, binary modifiers are often launched from the command line of the cmd utility in a similar fashion to Linux. The specific commands are shown below:

```
//Launching Pin on Windows (JIT mode)
pin.exe <pinargs> -t <pintool>.dll <pintoolargs> -- <app> <appargs>

//Launching DynamoRIO on Windows (JIT mode)
drrun.exe -client <client>.dll 0 "" <app> <appargs>
```

Meanwhile, some tools ship with graphical user interfaces that allow the user to launch applications using icons rather than command lines. These GUIs are generally intuitive to use (by design).

The above examples demonstrated the commands necessary to launch a system using JIT-mode execution. If the user wishes to use probe-mode execution instead, they must make some changes to their plug-in routines, and then the following commands can be invoked:

```
//Launching Pin (Probe mode)
pin -probe -t <pintool>.so <pintoolargs> -- <app> <appargs>
```

```
//Launching DynamoRIO on Windows (Probe mode)
drrun.exe -mode probe -client <client>.dll 0 "" <app> <appargs>
```

As can be inferred from all of the command lines above, the guest application is not actually launched at the command line, but it is simply passed as an argument to the dynamic binary modification system. The system will then initialize itself before launching the guest application under its control, modifying the application as dictated by the plug-in tool. (The means for acquiring and maintaining control are beyond the scope of this chapter and are discussed in Chapter 6 which focuses on the internal system implementation).

In addition to the arguments passed on the command line, many other arguments may be set via external files and/or environment variables. For instance, DynamoRIO accesses several environment variables including DYNAMORIO_OPTIONS where dozens of additional parameters may be set.

2.2.2   ATTACHING TO AN EXISTING PROCESS

Rather than running the entire execution time under the control of the dynamic binary modifier on some systems, it is also possible to attach the system to an already-running program, much like a debugger. This allows the application to run natively until the user performs the attach operation. The command below demonstrates using the *attach* model in Pin.

```
//Attaching to a running process in Pin
pin <pinargs> -t <pintool>.so <pintoolargs> -pid <app_pid>
```

Pin allows the user to *detach* from a program as well, allowing it to be run natively for the duration of the execution time. This can be useful for those ephemeral tasks where full coverage is unnecessary. Detaching is performed within the user-defined plug-in instrumentation tool by making an API call to PIN_Detach().

2.3   PROGRAMMABLE INSTRUMENTATION

Now that we've seen how to launch an application under the control of a binary modification system, it's important to understand how a user defines the modifications that they wish to apply to the running guest application.

One thing to note from the command lines used to invoke Pin, DynamoRIO, and Valgrind in the previous section was that some of the command-line arguments referred to a user-defined plug-in tool. On Pin, this was called a *pintool*, on DynamoRIO it was a *client*, and on Valgrind it was simply a *tool*. Regardless of the terminology, each dynamic binary modifier provides convenient abstractions and API's that a user can use to specify how and when to modify the guest application. These abstractions are then used together with standard code (often written in C or C++) to form a plug-in program that will be weaved into the running application by the dynamic binary modifier where it will run in the same address space as the guest application, as shown in Figure 2.1. Therefore, aside from understanding how to launch the system, a user must understand how to write a plug-in

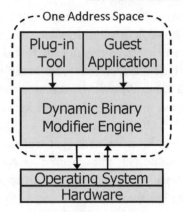

Figure 2.1: The binary modification engine, the guest application, and the user's plug-in tool all execute in the same address space.

specification for the changes they wish to make, which is really a matter of understanding the system's exported API.

Because these APIs differ greatly between system, we will focus on the Pin API for the purposes of this and the next few chapters, before moving back to a system-agnostic view when covering the internal implementations of dynamic binary modification systems in Chapter 6 and beyond. This chapter is by no means an extensive user manual defining the Pin API; the goal is to provide an intuitive sense of the power available to the user and the overall intent of the API. The interested reader is encouraged to visit the project website for each system to access the complete user guide.

API Overview   From the highest level, the API allows a user to iterate over the instructions that are about to execute, in order to have the opportunity to add, remove, change, or simply observe the instructions prior to executing them. The changes can be as simple as inserting instructions to gather dynamic profiling information, or as complex as replacing a sequence of instructions with an alternate implementation.

The most basic APIs provide common functionalities like determining instruction details, determining control-flow changes, or analyzing memory accesses. In Pin, most of the API routines are call-based. The user can register a *callback* to be notified when key events occur, and the user can make calls from their plug-in tool into the Pin engine to gather relevant information. (Note that in many cases, Pin will automatically inline these calls to improve performance, as will be discussed later. Meanwhile, some tools always assume that analysis code will be inlined, and they will leave it to the user to ensure that inlining is safe by saving and restoring any needed registers or state.)

Instrumentation vs. Analysis   At this point, it's important to provide a bit of terminology to distinguish the opportunities available for observing and modifying the guest application. Most systems provide two types of opportunities to observe the application – a static opportunity and a dynamic (runtime) opportunity. The static opportunity allows every distinct instruction that executes to be observed or modified once, and more specifically, the first time that instruction is seen. From that point forward, any of the static changes that were made to those instructions will persist for the duration of the execution time. We call the routines that provide this static opportunity *instrumentation code*. Instrumentation routines focus on specific code locations.

Alternatively, dynamic opportunities arise every time a single instruction executes at runtime. Measuring a dynamic event involves inserting code that will execute over and over for any given instruction. We call the routines that provide this dynamic view *analysis code*. Analysis code focuses on events that occur at some point within the execution of an application.

To summarize, instrumentation routines define *where* to insert instrumentation. Analysis routines define *what* to do when the instrumentation is activated. Instrumentation routines execute once per instruction. Analysis can execute millions of times for each instruction, depending on how deeply nested in loop code that one instruction lies. This terminology becomes particularly important when thinking about how to implement a desired goal. For instance, if the user wishes to gather the frequency of using a particular register, they will have to distinguish the static frequency (how often the register appears in the binary) from the dynamic frequency (how often the register is accessed at runtime). It is also important to distinguish these opportunities so that the user is not adding unnecessary dynamic overhead when some fixed amount of static overhead would suffice.

Instrumentation Points, Granularity, and Arguments   The system permits the user's plug-in tool to access every executed instruction. The tool can then choose to modify the particular instruction, or insert code before or after that instruction. For branch instructions, new code can be inserted on either the fall-through or taken path. The tool designer must be sure that the particular location they choose to insert code will actually execute. For instance, inserting new code after an unconditional jump is probably not a good idea.

The APIs of Pin, DynamoRIO, and Valgrind all allow the user or tool to iterate over and inspect several distinct granularities of the guest application. The user can choose to iterate over single instructions just before each instruction executes, entire basic blocks[2] (a straight-line sequence of non-control flow instructions followed by a single control flow instruction, such as a branch, jump, call, or return), entire traces[3] (a series of basic blocks), or the entire program image.

It is important to understand that the basic blocks and traces that the system presents to the tool represent a single-entry dynamic path. That is, control can only enter the top of the sequence (not the sides), but control can exit through any side exit that exists. If later, control enters the side of an existing sequence, then a new structure (basic block or trace) will be formed starting at the side entry point. Therefore, there can be duplication between basic blocks and traces when performing static analysis! This fact is important for users to understand if their results depend on all

---

[2]Technically, what is commonly called a basic block in the dynamic binary modification world is actually an extended basic block in the literature, as the system cannot tell whether any of the straight-line instructions are targets of other branches.

[3]What is called a trace in the dynamic binary modification world is called a superblock in the static compiler world.

Table 2.1: Common Dynamic Binary Modification Systems. The table lists the supported architectures, supported operating systems, operating modes, and websites of each system.

System Name	Architectures	OSes	Features
Pin (pintool.org)	x86,x86-64, Itanium,ARM	Linux,Windows, MacOS	JIT,Probe, Attach
DynamoRIO (dynamorio.org)	x86,x86-64	Linux,Windows	JIT,Probe
Valgrind (valgrind.org)	x86,x86-64, ppc32,ppc64,ARM	Linux,MacOS	JIT

statically-reported instruction sequences being distinct, such as gathering a static instruction count. In practice, the duplication affects a very small proportion of instructions. For situations where this matters, the user can distinguish unique instances of static instructions using the original memory address of an instruction as an indicator of uniqueness, rather than assuming that all instructions reported at instrumentation time are unique.

2.4   PLATFORM-SPECIFIC BEHAVIORS

The Pin API (like that of other systems) provides both architecture-independent and architecture-dependent abstractions about the running application's instructions. The most basic APIs provide common functionalities like determining control-flow changes or memory accesses. For instance, the boolean API `IsBranch()` for determining whether a given instruction can potentially change the control flow is architecture independent from the user's perspective. Clearly, the internal implementation must use an architecture-specific decoder to determine whether the instruction is a branch, but this complexity is masked from the user. Meanwhile, when a user wants to access information about specific opcodes or operands, such as whether the target operand in an instruction matches the register `%eax`, they may then access the architecture-specific APIs. The idea behind most APIs is to hide the unnecessary complexity while still providing the power to perform essential tasks and analyses.

Many dynamic binary modifiers support more than one operating system. For instance, Pin and DynamoRIO support both Windows and Linux. For the most part, the system developers have designed the API to abstract away the differences between operating systems, and the main implementation differences occur behind the scenes. There are platform-specific APIs, such as the `PIN_GetWindowsExceptionCode()` routine, for instances when it makes sense to expose the details of the particular operating system. See Table 2.1 for the various platforms supported by the three systems discussed in this chapter. Note that not all architecture/OS combinations listed exist for each system. For instance, Pin for ARM only supports Linux, not Windows or MacOS (Hazelwood and Klauser [2006]).

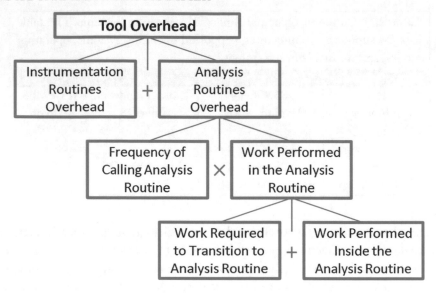

Figure 2.2: Understanding the Sources of Overhead in Your Plug-In Tool.

2.5 END-USER OPTIMIZATIONS

Pin and other dynamic binary modifiers will analyze the code changes requested by the user and will automatically handle all of the necessary setup code required. For instance, if the user wishes to insert a new profiling instruction, that instruction will undoubtedly require registers to operate. However, the user should not have to analyze the surrounding code to determine whether free registers exist. Instead, the system will perform this task; it will determine whether free registers exist, and it will generate any necessary spill and fill code if necessary. On systems such as Pin, any new functions generated by the user will be analyzed to determine whether they can be inlined into the guest application code, or whether they are best left as function calls. Finally, all other preparation work for ensuring that the guest application will otherwise appear to behave identically to a native run is handled automatically by the system. These tasks all carry overhead but are generally not responsible for the bulk of the overhead experienced by users.

As discussed in Section 1.3, the overhead experienced by the user of a dynamic binary modifier depends on a number of factors. The baseline overhead of most systems is quite low, so the largest contributor to the overhead is that of the user-defined tool itself, which is in many ways manageable by the informed user.

Figure 2.2 shows the breakdown of contributors to the overhead of the user's tool. The bulk of the overhead comes from the analysis routines, as they execute much more frequently than the static instrumentation routines. Therefore, if any of the work can be relocated from analysis to instrumentation routines, the user will notice significant speedups. For instance, address calculations or data structure locations may not change dynamically, and if this is the case, the calculations should be performed once, in the instrumentation routine, and then the calculated address can be passed directly to the analysis routine (rather than recalculating it upon every dynamic occurrence). From there, if the user can reduce the frequency at which the analysis routines are called, this will reduce overhead as well. This can often be accomplished by operating at larger granularities of the application's code. For instance, counting instructions can safely be performed one instruction at a time, or one basic block at a time, but the latter will have significantly lower overhead.

Finally, there are a few tricks that the user can perform to reduce the amount of work done inside the analysis routine if they understand that part of that overhead comes in transitioning to the analysis routine and if they understand what choices can affect that transition overhead. For example, as mentioned earlier, systems like Pin will analyze any new analysis code to determine whether it can be inlined directly into the surrounding instructions or whether it should be converted to a function call. The main metric Pin uses for making this decision is the simplicity of the new code, and specifically, whether or not control-flow changes or calls exist within the new code. Therefore, any conditional code is better implemented by escalating the condition checking routine to instrumentation time, if possible, rather than checking the condition inside every dynamic instance. Pin provides some special conditional instrumentation routines that the user can leverage to aid the system to perform inlining in the presence of simple conditional statements. Meanwhile, systems like Valgrind will allow the tool author to either inline analysis code or to generate calls to separate analysis functions (written in C).

2.6   DEBUGGING YOUR PLUG-IN TOOL

Assuming a correctly implemented guest application, and a correctly implemented binary modification engine, the final task that a user of a binary modifier is likely to require is a way to debug their custom plug-in tool. The fact that three applications are actually running in the same address space (the binary modifier, the guest application, and the user's plug-in tool) means that standard debugging methodologies will not apply. Instead, documentation for each system provides specific details about the best way to debug plug-in tools on that system.

On Pin, for example, it is possible to use gdb to debug a user plug-in tool on Linux. However, the process involves using two different shells, one to run the debugger, and one to run the application under the control of Pin. The three step process is shown below:

Step 1   In one window, invoke gdb with Pin:

```
prompt% gdb pin
(gdb)
```

Step 2   In a second window, launch your Pintool with the -pause_tool flag, which takes the number of seconds to pause as an argument.

```
prompt% pin -pause_tool 5 -t myPinTool.so -- <guestApp>
Pausing to attach to pid 32017
```

Step 3   Back in the gdb window, attach to the paused process. You may now use gdb in the standard fashion, setting breakpoints as usual, and running cont to continue execution.

```
(gdb) attach 32017
(gdb) break main
(gdb) cont
```

Other systems will have their own tricks for debugging the variety of execution modes on the variety of supported platforms in their user manuals.

Summary   At this point, we now understand the high-level applications and necessary terminology for using dynamic binary modification systems. We will therefore move on to specific examples and use cases in the following chapters.

CHAPTER 3

Program Analysis and Debugging

A fundamental motivation for developing dynamic binary modification systems, initially, was that they would provide a simple yet robust way to instrument and analyze the behavior and bottlenecks of guest programs. In fact, a variety of industrial products have been developed that use dynamic binary modification as the technological foundation. For instance, the Pin dynamic instrumentation system forms the base of a number of Intel products, including Parallel Inspector, Parallel Amplifier, which are parallel program analysis tools, and TraceCollector and SDE, which are simulation and emulation tools. In addition, DynamoRIO formed the foundation of the Determina Vulnerability Protection Suite, which is a set of security tools.

Therefore, this chapter provides some concrete examples of the application of dynamic binary modification to the area of program analysis. We incorporate several sample applications from the Pin sample plug-in tool library, and we analyze the source code and outputs from each example. These sample tools can easily form the foundation for similar yet more sophisticated tools.

3.1 PROGRAM ANALYSIS EXAMPLES

In this section, we will cover four simple program analysis Pintools that demonstrate the ease of analyzing a running program using dynamic binary modification.

Generating a Dynamic Instruction Trace with `PrintPC` Perhaps one of the simplest plug-in tools that can be written is one that performs the task of printing the machine's program counter throughout the execution of an application. Such a tool can be useful for gathering a statistical view of where the execution time is spent. Figure 3.1 demonstrates the entire program necessary to implement this functionality as a plug-in to the Pin dynamic instrumentation system, and it demonstrates some of the basic APIs provided to the user.

The easiest way to understand `PrintPC` is to start from the bottom of Figure 3.1 and focus on the `main()` routine. Here we see that the user makes some calls to initialize their data structures, open any necessary output files, and initialize Pin. Next, they register an instrumentation routine that will be executed for every static instruction seen at runtime (line 26). Finally, they register another routine that will execute immediately prior to exiting at the end of the execution time (line 27), before instructing Pin to launch the guest application (line 28). Nothing after the call to `PIN_StartProgram()` will ever execute. (The call to `return 0` is only present to make the compiler happy.)

```
                                        ─── The PrintPC Tool ───
 1  ofstream TraceFile;
 2
 3  // This analysis call is invoked for every dynamic instruction executed
 4  VOID PrintPC(VOID *ip)
 5  {
 6      TraceFile << ip << endl;
 7  }
 8
 9  // This instrumentation routine is invoked once per static instruction
10  VOID Instruction(INS ins, VOID *v)
11  {
12      INS_InsertCall(ins, IPOINT_BEFORE, (AFUNPTR)PrintPC, IARG_INST_PTR, IARG_END);
13  }
14
15  // This fini routine is called after the guest program terminates, just prior to exiting Pin
16  VOID Fini(INT32 code, VOID *v)
17  {
18      TraceFile.close();
19  }
20
21  // This main routine is invoked to initialize Pin, prior to starting the guest application
22  int main(int argc, char * argv[])
23  {
24      TraceFile.open("pctrace.out");          // Open an output file
25      PIN_Init(argc, argv);                   // Initialize Pin
26      INS_AddInstrumentFunction(Instruction, 0);  // Register a routine to be called to instrument instructions
27      PIN_AddFiniFunction(Fini, 0);           // Register Fini to be called when the application exits
28      PIN_StartProgram();                     // Start the program; this call never returns
29      return 0;
30  }
```

Figure 3.1: This program analysis tool prints the address of every instruction that executes to a file. It demonstrates the use of static instruction-level instrumentation routines and dynamic analysis routines.

Next, let's look at the instrumentation and analysis routines. The instrumentation routine is called `Instruction()` (line 10), and it will be called every time an instruction is encountered for the first time. When that occurs, we tell the system to insert a new routine before that instruction, which will be called (immediately prior to) when the instruction executes (line 13). The new routine we insert before every instruction is called `PrintPC()` (line 4). It takes the current instruction pointer (program counter) as an argument, then prints that PC to a file.

If we compile `PrintPC`, link it to Pin's libraries, and execute a guest application using this plug-in, the output will be a (large) trace file. The file will contain a list of program addresses that executed, in the order that they executed, including all executed addresses within shared libraries. Since the plug-in will run in user space alongside the guest application, no kernel addresses will appear in the trace file.

This simple tool can easily be extended to sample the program counter, rather than to print every single address. It can further be optimized to use conditional instrumentation to reduce the overhead of PC sampling. Finally, it can be extended to print not only the instruction addresses, but

─────────────────── The CallTrace Tool ───────────────────

```
1  // One of the two following analysis routines will be invoked for every dynamic call
2  VOID  do_call(const string *s)
3  {
4      TraceFile << *s << endl;
5  }
6  VOID  do_call_indirect(ADDRINT target, BOOL taken)
7  {
8      if( !taken ) return;
9      do_call( Target2String(target) );
10 }
11
12 // This instrumentation routine is invoked once per static trace
13 VOID Trace(TRACE trace, VOID *v)
14 {
15     for (BBL bbl = TRACE_BblHead(trace); BBL_Valid(bbl); bbl = BBL_Next(bbl)) {
16         INS tail = BBL_InsTail(bbl);
17         if( INS_IsCall(tail) ) {
18             if( INS_IsDirectBranchOrCall(tail) ) {
19                 const ADDRINT target = INS_DirectBranchOrCallTargetAddress(tail);
20                 INS_InsertPredicatedCall(tail, IPOINT_BEFORE, AFUNPTR(do_call),
21                                          IARG_PTR, Target2String(target), IARG_END);
22             }
23             else INS_InsertCall(tail, IPOINT_BEFORE, AFUNPTR(do_call_indirect),
24                                 IARG_BRANCH_TARGET_ADDR, IARG_BRANCH_TAKEN, IARG_END);
25         }
26     }
27 }
28
29 // This fini routine is called after the guest program terminates, just prior to exiting Pin
30 VOID Fini(INT32 code, VOID *v)
31 {
32     TraceFile.close();
33 }
34
35 // This main routine is invoked to initialize Pin, prior to starting the guest application
36 int  main(int argc, char *argv[])
37 {
38     PIN_InitSymbols();
39     PIN_Init(argc,argv);
40     TraceFile.open();                       // Opens the output file
41     TRACE_AddInstrumentFunction(Trace, 0);  // Will gather the call trace as the program runs
42     PIN_AddFiniFunction(Fini, 0);           // Closes the output file prior to exiting
43     PIN_StartProgram();                     // Launches the program and never returns
44     return 0;
45 }
```

Figure 3.2: The CallTrace program analysis tool records all function calls that occur during execution. This is a simplified version of the tool available in SimpleExamples/calltrace.cpp of the Pin distribution, originally written by Robert Muth.

the actual instructions, including opcodes and operand values. We leave these tasks as an exercise for the interested reader.

Call-Graph Generation with `CallTrace`   Another simple profiling tool is one that analyzes the function calls made while a program runs. Such a tool could be used for analyzing code coverage or for detecting inefficiencies in the call stream. Figure 3.2 demonstrates a simple call trace generation tool. Unlike the previous example, this tool instruments entire traces at once, rather than individual instructions. Working at larger granularities makes for a more efficient design that has a lower run-time overhead. We see a difference in the implementation first on line 43, where we use the `TRACE` API rather than the `INS` API. Next, within the instrumentation routine `Trace()` on line 13, we iterate over all instructions within the trace to search for a call, rather than handling one instruction at a time. We also see a few more query APIs that are available, such as the boolean `INS_IsCall()` query on line 19 and the `INS_IsDirectBranchOrCall()` query on line 20. This allows the tool to distinguish between direct and indirect calls statically, and to insert the corresponding specialized analysis calls for each case. This also demonstrates a subtle point, which is that any query that can be done once statically should be done at that time. While we could have embedded the query function for determining whether an instruction is a branch or call into the dynamic stream, this trait does not change at runtime, and it would therefore be inefficient to query the same instruction multiple times.

If we execute `CallTrace` as a plug-in to Pin while running a guest application, we will get a log of all of the function calls that were made (by name) throughout execution. Again, we will see only user-level behavior, which includes calls to shared and dynamically-loaded libraries and even system calls, but no calls made from within the kernel. We can also extend the CallTrace example in a number of ways. We can modify the tool to print the arguments to each call, to focus exclusively on system calls, or to focus on one particular call, such as `malloc()`. Finally, we can write a more sophisticated tool to generate a call graph, rather than simply a log of all calls.

Memory-Leak Detection with `MallocTrace`   Rather than focusing on all dynamic calls, a natural extension is focus in on a few calls of interest, such as those relating to memory allocations. Such a tool can be used to easily detect memory leaks within applications by comparing the amount of memory allocated to the amount deallocated.

The tool shown in Figure 3.3 shows a simple way to modify one or more particular functions of interest. In this case, the tool instruments any application call to `malloc()` or `free()` by modifying the contents of the functions themselves. We accomplish this by performing instrumentation on the entire image at load time, as is demonstrated on line 39 of the tool. The instrumentation routine itself searches for the two routines of interest (the `malloc()` and `free()` routines) on lines 17 and 25, respectively. This search is performed once. If either routine is located, the system inserts the new functionality shown in the two analysis routines called `BeforeMallocFree()` and `AfterMalloc()`, which simply prints some information about the size and location of the allocation or deallocation by analyzing the inputs to the calls themselves from within the function body. Since the details of each call to `malloc()` and `free()` will vary at run-time as different amounts of memory are requested or released, we must track the arguments and return values to/from each call. We accomplish this in Pin by specifying a set of arguments that can be captured at runtime and passed to the plug-in analysis routines. The arguments themselves are specified on lines 20-21, 22, and 28-29. The arguments of interest are already captured by the Pin API, which allows the user to access the `FUNCARG_ENTRYPOINT_VALUE` (the inputs to a function) and/or the `FUNCRET_EXITPOINT_VALUE` (the return value from a function). These values are then passed to the analysis routines to be printed

——————— The MallocTrace Tool ———————

```
1  #define MALLOC "malloc"
2  #define FREE "free"
3
4  // The following analysis calls are invoked before/after every call to malloc and free
5  VOID BeforeMallocFree(CHAR * name, ADDRINT size)
6  {
7      cout << name << "(" << size << ")" << endl;
8  }
9  VOID AfterMalloc(ADDRINT ret)
10 {
11     cout << "  returns " << ret << endl;
12 }
13
14 // This image routine is invoked once, prior to executing the program
15 VOID Image(IMG img, VOID *v)
16 {
17     RTN mallocRtn = RTN_FindByName(img, MALLOC);  // Finds malloc()
18     if (RTN_Valid(mallocRtn)) {
19         RTN_Open(mallocRtn);
20         RTN_InsertCall(mallocRtn, IPOINT_BEFORE, (AFUNPTR)BeforeMallocFree, IARG_ADDRINT, MALLOC,
21                        IARG_FUNCARG_ENTRYPOINT_VALUE, 0, IARG_END);
22         RTN_InsertCall(mallocRtn, IPOINT_AFTER, (AFUNPTR)AfterMalloc, IARG_FUNCRET_EXITPOINT_VALUE, IARG_END);
23         RTN_Close(mallocRtn);
24     }
25     RTN freeRtn = RTN_FindByName(img, FREE);     // Finds free()
26     if (RTN_Valid(freeRtn)) {
27         RTN_Open(freeRtn);
28         RTN_InsertCall(freeRtn, IPOINT_BEFORE, (AFUNPTR)BeforeMallocFree, IARG_ADDRINT, FREE,
29                        IARG_FUNCARG_ENTRYPOINT_VALUE, 0, IARG_END);
30         RTN_Close(freeRtn);
31     }
32 }
33
34 // This main routine is invoked to initialize Pin, prior to starting the guest application
35 int main(int argc, char *argv[])
36 {
37     PIN_InitSymbols();                    // Initialize Pin's symbols for RTN instrumentation
38     PIN_Init(argc,argv);                  // Initialize Pin
39     IMG_AddInstrumentFunction(Image, 0);  // Register a routine to be called to instrument the image
40     PIN_StartProgram();                   // Start the program; this call never returns
41     return 0;
42 }
```

Figure 3.3: The MallocTrace program analysis tool instruments the `malloc()` and `free()` functions. It prints the arguments to each function, and the return value from `malloc()`.

out a runtime. The net result of applying the `MallocTrace` tool is that we have an application that, rather than calling the native `malloc` and `free` routines, will instead call a new version of `malloc` and `free`. The new versions are, otherwise, identical to the old, but they will be amended to contain new code that prints the arguments to these routines and the return values. This corresponding log of memory allocations and deallocations can subsequently be analyzed to detect memory leaks. While the log itself is generated as the guest application executes, the task of detecting memory leaks can be performed either online during execution, or offline after the guest application completes, and the log has been written to a file. These are good starting points for realistic memory leak tools.

```
                              ── The InsMix Tool ──
 1  // This analysis call is invoked for every dynamic instruction executed
 2  VOID PIN_FAST_ANALYSIS_CALL docount(COUNTER * counter)
 3  {
 4      (*counter)++;
 5  }
 6
 7  // This instrumentation routine is invoked once per static trace. It inserts the analysis routine.
 8  VOID Trace(TRACE trace, VOID *v)
 9  {
10      for (BBL bbl = TRACE_BblHead(trace); BBL_Valid(bbl); bbl = BBL_Next(bbl)) {
11          // Insert instrumentation to count the number of times the bbl is executed
12          BBLSTATS * bblstats = new BBLSTATS(stats, INS_Address(BBL_InsHead(bbl)), rtn_num, size, numins );
13          INS_InsertCall(BBL_InsHead(bbl), IPOINT_BEFORE, AFUNPTR(docount), IARG_FAST_ANALYSIS_CALL, IARG_PTR,
14                         &(bblstats->_counter), IARG_END);
15      }
16  }
17
18  // This image routine is invoked once, prior to executing the program
19  VOID Image(IMG img, VOID * v)
20  {
21      for (SEC sec = IMG_SecHead(img); SEC_Valid(sec); sec = SEC_Next(sec)) {
22          for (RTN rtn = SEC_RtnHead(sec); RTN_Valid(rtn); rtn = RTN_Next(rtn)) {
23              // A RTN is not broken up into BBLs, it is merely a sequence of INSs
24              RTN_Open(rtn);
25              for (INS ins = RTN_InsHead(rtn); INS_Valid(ins); ins = INS_Next(ins)) {
26                  for(UINT16 *start=array; start<end; start++) GlobalStatsStatic[ *start ]++;
27              }
28              RTN_Close(rtn); // to preserve space, release data associated with RTN after processing
29          }
30      }
31  }
```

Figure 3.4: Snippet of the InsMix program analysis tool, which categorizes the static and dynamic instruction stream. The complete tool is available in Insmix/insmix.cpp of the Pin distribution and was originally written by Robert Muth.

Instruction Profiling and Code Coverage with `InsMix`   Our final program analysis example focuses on the general case of instruction profiling. There are a number of reasons that developers may wish to profile the static and dynamic instruction stream of their applications. The `InsMix` tool shown in Figure 3.4 serves as the foundation for a number of such profiling tools. This tool can be used to group instructions by class to determine the most frequently used and most frequently executed instruction classes for a given program. Such a tool can be used for code coverage analysis, which can be done by comparing the static and dynamic instruction stream. Interestingly, this same tool can be used for compiler bug detection. By comparing the code generated by one compiler to that generated by another compiler, we can easily detect inefficiencies in the code generation routines, such as unnecessary spills and fills.

3.2   PARALLEL PROGRAM ANALYSIS

Developers can use dynamic binary modification to analyze parallel program performance and correctness. Many tools can be written to find memory and threading errors, such as memory leaks, references to uninitialized data, data races, and deadlocks. These tools can use the binary modification engine to instrument the running program and collect the information necessary to detect these errors.

For instance, a *data race* occurs when two threads access the same data, at least one access is a write, and there is no synchronization (for example, locking) between accesses (see Banerjee et al. [2006]). Unsynchronized variable writes usually are a programming error and can cause nondeterministic behavior. To detect data races, the Intel Parallel Inspector product uses Pin to instrument all machine instructions in the program that reference memory and records the effective addresses. It also instruments calls to thread synchronization APIs. By examining the effective addresses, it is possible to detect when multiple threads access the same data. Meanwhile, by instrumenting calls to thread synchronization APIs, it is possible to determine whether the memory accesses were synchronized. Finally, to help the programmer identify the cause of the data race, the system APIs can be used to trace back to source code lines leading to the problematic memory references.

Another useful parallel program analysis tool is called `LocksAndWaits`, and it is part of the Intel Parallel Amplifier toolset. LocksAndWaits measures the time multithreaded programs spend waiting on locks, attributing time to synchronization objects and source lines. Identifying locks responsible for wait time and the associated source lines helps programmers improve a parallel program's CPU utilization. The LocksAndWaits analysis uses Pin's probe mode to replace calls to synchronization APIs with wrapper functions. The wrapper functions call the original synchronization function and record the wait time, synchronization object, and call stack.

Even state-of-the-art compilers miss many parallelization opportunities in C/C++ programs, and as a result, programmers are forced to manually parallelize applications. The success of manual parallelization relies on execution profiler quality. Unfortunately, popular execution profilers operate at function or instruction granularity, which is insufficient for parallel programming because many programs are parallelized at the loop level. Therefore, developers created the Pin-based Prospector tool which discovers potential parallelism in serial programs by profiling loops and data dependences. Prospector provides information such as loop trip counts and the number of instructions executed inside loops. It also dynamically detects loop-carried data dependencies, which must be preserved during the parallelization process. Programmers then receive reports on candidate loops for parallelization. In addition to the profiler, Prospector provides several tools for visualizing and interpreting the profiling results.

A gentle introduction to the potential applications of one dynamic binary modifier to the area of parallel program analysis is presented by Bach et al. [2010], while more details on Prospector is presented by Kim et al. [2009].

3.3   DETERMINISTIC REPLAY

Debugging and analyzing parallel programs is difficult because their execution is not deterministic. The threads' relative progress can change in every run of the program, possibly changing the results.

Even single-threaded program execution is not deterministic because of behavior changes in certain system calls (for example, `gettimeofday()`) and stack and shared library load locations.

Using dynamic binary modification, it is possible to perform user-level capture and deterministic replay of multithreaded programs. PinPlay is one such tool, based on Pin. The program first runs under the control of a logging tool, which captures all the system call side effects and inter-thread shared-memory dependencies. Another tool replays the log, exactly reproducing the recorded execution by loading system call side effects and possibly delaying threads to satisfy recorded shared-memory dependencies.

Replaying a previously captured log by itself is not very useful. Instead, a captured log can be used to ensure that other program analysis tools see the same program behavior on multiple runs, making the analysis deterministic. The tool can also replay a PinPlay log while connected to a debugger, making multithreaded program debugging deterministic. As long as the PinPlay logger can capture a bug once, the behavior can be precisely replicated multiple times with replay under a debugger. More details of PinPlay are presented by Patil et al. [2010], while more details on the logging operating system effects is presented by Narayanasamy et al. [2006].

3.4   CUSTOMIZABLE DEBUGGING

Given that debuggers like `gdb` are used for interactively querying program state, combining a dynamic binary modifier with a debugger can provide powerful opportunities to extend the functionality of the debugger itself, or often even provide debugging capabilities at significantly improved speeds. Yet, combining these tools can be a challenge without explicit support since the debugger expects to see unaltered program state while the binary modifier will alter that state significantly. Therefore, systems such as Pin have built-in support for extensible, customizable debugging through an advanced debugging interface. This interface enables seamless integration with off-the-shelf debuggers on Windows and Linux by providing the illusion of unmodified guest program state that the debugger expects. Meanwhile those debuggers can then operate with new, customizable, user-defined functionalities.

One example of new functionality that a user may wish to implement is the ability to use a standard debugger to debug applications that feature new emulated instructions or explore the use of new registers. Users have combined the Intel Software Development Emulator (SDE) with the Pin Advanced Debugger Extensions (PinADX) to debug emulated programs as if they are debugging them natively, where new instructions and registers are seamlessly emulated and visible to the debugger.

Another example of integrating a dynamic binary modifier with a standard debugger has been done in the context of deterministic record-and-replay. A multi-threaded program's execution with a thread-order specific bug can be recorded with PinPlay's logger tool while PinPlay's replayer tool can repeat the buggy behavior as many times as needed. When PinADX is combined with the PinPlay replayer tool, the user can perform a series of debugger sessions with a guarantee of replaying the bug each time, improving the ease of isolating its cause.

CHAPTER 4

Active Program Modification

While the previous chapter showed several examples of using dynamic binary modification systems to analyze or debug programs, all such examples were passive in that they inserted new code to observe behaviors of the guest application, but not modify those behaviors. By contrast, this chapter focuses on tools that actively modify the original functionality of an application. Modifications can be fine grained, such as changing register or memory values of individual instructions, or modifying the control flow of a program. They can also be coarse grained, such as adding or deleting relevant functionality, replacing entire procedures, or applying optimizations or security features to an entire guest application.

4.1   FINE-GRAINED INSTRUCTION MODIFICATION

Dynamic binary modifiers allow for the instructions of the guest application to be altered in a variety of ways. Entire instructions can be inserted or deleted, or various operands can be independently modified within existing instructions, such as register values, memory addresses, or control flow. Some examples of the pertinent APIs in the Pin system that allow individual instructions to be modified include the following:

```
// Deleting an Arbitrary Instruction
void INS_Delete (INS ins)

// Inserting New Control-Flow Instructions
void INS_InsertDirectJump (INS ins, IPOINT ipoint, ADDRINT tgt)
void INS_InsertIndirectJump (INS ins, IPOINT ipoint, REG reg)

// Modifying the Operands of Existing Instructions
void INS_RewriteMemoryOperand (INS ins, UINT32 memopIdx, REG newBase)
```

To demonstrate the use of one such API in a real scenario, Figure 4.1 shows a snippet of a tool that locates instructions containing operands that read from or write to memory. The tool iterates through these memory operands on line 4, and systematically rewrites them to reference a new virtual memory location (where the new location is contained in the register specified on line 10).

Through the use of these and the other APIs available in the various binary modification systems, a user can essentially make any arbitrary fine-grained modification they can imagine. The APIs provide the flexibility to decide whether to modify an instruction in place or to delete the instruction and create one or more completely new instructions in its place.

```
                              ──────── Rewriting Memory Operands ────────
 1  // Map the original effective address (originalEa) to a translated address
 2  static ADDRINT ProcessAddress(ADDRINT originalEa, ADDRINT size, UINT32 access);
 3  ...
 4      for (UINT32 op = 0; op<INS_MemoryOperandCount(ins); op++) {
 5          UINT32 access = (INS_MemoryOperandIsRead(ins,op)    ? 1 : 0) |
 6                          (INS_MemoryOperandIsWritten(ins,op) ? 2 : 0);
 7          INS_InsertCall(ins, IPOINT_BEFORE, AFUNPTR(ProcessAddress),
 8                         IARG_MEMORYOP_EA, op, IARG_MEMORYOP_SIZE, op,
 9                         IARG_UINT32, access, IARG_RETURN_REGS, REG_INST_G0+i, IARG_END);
10          INS_RewriteMemoryOperand(ins, i, REG(REG_INST_G0+i));
11      }
12  ...
```

Figure 4.1: Snippet of a tool for rewriting the memory operations in a guest application.

4.2   FUNCTION REPLACEMENT

Often, the desired program modification may be much more coarse grained, such that inserting, modifying, or deleting individual instructions would be far too tedious. It is fairly straightforward to replace an entire routine in a guest application or shared library with a custom routine even without access to the source code of that routine. This type of function replacement is a typical example of active program modifications that are possible in a dynamic binary modification system.

```
                              ──────── The ReplaceMalloc Tool ────────
 1  void MyNewMalloc()
 2  {
 3      // Get a handle to the original malloc so we can call it
 4      typeof(MyNewMalloc) * original = (typeof(MyNewMalloc)*)PIN_RoutineWithoutReplacement();
 5
 6      cerr << "In replacement" << endl;
 7      original();
 8      cerr << "After replacement" << endl;
 9  }
10
11  VOID ImageLoad(IMG img, VOID *v)
12  {
13      RTN mallocRtn = RTN_FindByName(img, "malloc");
14      if (RTN_Valid(mallocRtn))
15      {
16          RTN_ReplaceProbed(mallocRtn, AFUNPTR(MyNewMalloc));
17          cout << "Inserted probe for malloc:" << endl;
18      }
19  }
20
21  int main(int argc, CHAR *argv[])
22  {
23      PIN_InitSymbols();
24      PIN_Init(argc,argv);
25      IMG_AddInstrumentFunction(ImageLoad, 0);
26      PIN_StartProgramProbed();
27      return 0;
28  }
```

Figure 4.2: This tool replaces calls to `malloc()` with calls to a custom memory allocation routine.

Figure 4.2 shows an example of function replacement. In this case, the user would like to replace all calls to `malloc()` on Linux with a call to a custom memory allocation routine. The custom routine, called `MyNewMalloc`, begins on Line 1. Note that, in this case, the custom routine simply acts as a wrapper function that calls the standard `malloc()`, routine after printing a custom message. Yet, there is no restriction in place that a new routine must call the routine it has replaced, so a truly custom memory allocator could be implemented and deployed, and in fact, this example code would serve as a solid template for doing so.

The function replacement plug-in shown in Figure 4.2 also demonstrates two key features of dynamic binary modification systems. First, when deployed, the tool will replace *all* calls to `malloc()`, including those made by the shared libraries invoked by the application – not just the calls contained within the guest application itself. Second, the tool demonstrates the use of *probe-based instrumentation* where the calls to malloc are overwritten prior to executing the application, at load time, rather than on-the-fly as the calls are executed.

4.3 DYNAMIC OPTIMIZATION

Some of the earlier dynamic binary modifiers were designed to perform optimizations on running applications with the goal of a net performance improvement. One example system was Dynamo from Hewlett-Packard (Bala et al. [1999, 2000]), which optimized PA-RISC applications running on the HPUX operating system. Dynamo applied a series of optimizations designed to leverage the fact that at runtime, the program inputs are known; example optimizations included constant propagation and dead-code elimination. Since the goal of Dynamo was optimization, the system had the ability to "bail out" and execute the application natively if the act of binary modification was seen to be causing a performance degradation. Most of the modern dynamic binary modifiers do not support this feature as they are intended to provide comprehensive control over a guest application.

The DynamoRIO project from MIT spawned out of the original Dynamo project at HP (albeit ported to x86 on Linux and Windows rather than PA-RISC on HPUX), so DynamoRIO's goal has been one of dynamic optimization from the very beginning. Many of the design decisions within DynamoRIO reflect this emphasis, as DynamoRIO provides for fine-grained control over the internal behavior of its code generation engine and therefore the resulting performance of a modified guest application. Given this fact, it seems logical that our example demonstration of dynamic optimization will use DynamoRIO as the binary modification engine.

Figure 4.3 presents one of the standard demonstrations of dynamic optimization within DynamoRIO. The optimization leverages a processor-specific feature that would otherwise be too processor specific to implement in a static optimization framework. The optimization developer made the keen observation that when performing x86 assembly instruction selection for the high-level language statement `i++`, there are two choices which perform differently on two different processors. Specifically, the assembly instruction `inc` (for *increment*) is faster on the Pentium-III processor, while the `addi` (for *add immediate* where the constant operand is the value 1) is faster on

─────────────── **The IncVsAdd DynamoRIO Client** ───────────────

```
1  EXPORT void dr_init() {
2    if (proc_get_family() == FAMILY_PENTIUM_IV) dr_register_trace_event(event_trace);
3  }
4  static void event_trace(void *drcontext, app_pc tag, instrlist_t *trace, bool x18) {
5    instr_t *instr, *next_instr; int opcode;
6    for (instr = instrlist_first(bb); instr != NULL; instr = next_instr) {
7      next_instr = instr_get_next(instr);
8      opcode = instr_get_opcode(instr);
9      if (opcode == OP_inc || opcode == OP_dec) replace_inc_with_add(drcontext, instr, trace);
10   }
11 }
12 static bool replace_inc_with_add(void *drcontext, instr_t *instr, instrlist_t *trace) {
13   instr_t *in; uint eflags; int opcode = instr_get_opcode(instr);
14   bool ok_to_replace = false;
15   for (in = instr; in != NULL; in = instr_get_next(in)) {
16     eflags = instr_get_arith_flags(in);
17     if ((eflags & EFLAGS_READ_CF) != 0) return false;
18     if ((eflags & EFLAGS_WRITE_CF) != 0) {
19       ok_to_replace = true;
20       break;
21     }
22     if (instr_is_exit_cti(in)) return false;
23   }
24   if (!ok_to_replace) return false;
25   if (opcode == OP_inc) in = INSTR_CREATE_add (drcontext, instr_get_dst(instr, 0), OPND_CREATE_INT8(1));
26   else in = INSTR_CREATE_sub(drcontext, instr_get_dst(instr, 0), OPND_CREATE_INT8(1));
27   instr_set_prefixes(in, instr_get_prefixes(instr));
28   instrlist_replace(trace, instr, in);
29   instr_destroy(drcontext, instr);
30   return true;
31 }
```

Figure 4.3: This DynamoRIO client tool determines whether the underlying processor is Pentium-IV, and if so, it replaces each increment instruction with an add instruction dynamically.

the Pentium-IV. Therefore, the client tool determines the specific underlying processor at runtime, and dynamically converts the instruction selections when appropriate.

Digging deeper into the source code shown in Figure 4.3, we see that the client tool begins by registering an event that will be invoked upon initialization if the underlying processor is a Pentium-IV (shown on lines 1–3). Next, the system walks through the various instructions as they are encountered, searching for instances of the increment or decrement instructions (line 9). When found, the system invokes the routine `replace_inc_with_add` on that instruction. This particular routine is then shown on lines 12–31, where it begins by determining whether the instruction replacement is safe to perform. Next, it creates a new instruction that adds or subtracts the immediate value 1, and inserts it into the instruction stream before deleting the existing increment or decrement instruction. For the remainder of the current execution, the modified code sequence will persist, and, therefore, the act of replacing the instruction only occurs once per static instance of the increment or decrement instruction.

Many additional dynamic optimizations can also be envisioned and explored using dynamic binary modification systems. The ideal optimization candidates are those that cannot be performed statically because they are either too aggressive, too processor specific, or input specific, but that can be safely applied once the runtime environment is known.

4.4   SANDBOXING AND SECURITY ENFORCEMENT

A number of research groups have noted and leveraged the potential of dynamic binary modification for enforcing runtime security features on existing applications. The fact that the binary modification software layer can essentially act as a sandbox, allowing the system to observe potentially malicious or atypical behavior and actively prevent any repercussions on the running applications, is a powerful opportunity.

Perhaps the earliest work to note and leverage the security application was the work termed *program shepherding* by Kiriansky et al. (Kiriansky et al. [2002]) from MIT. The authors observe the control-flow transfers in running applications and subsequently apply three runtime security policies to prevent execution of malicious code (rather than preventing the code from being generated). First, they restrict execution privileges on code that is located on the stack, to prevent execution of malicious code masquerading as data. Second, they leverage the fact that dynamic binary modification systems observe not only application code, but also shared library code, by ensuring that shared library code is only entered at known entry points. Finally, they guarantee robust sandboxing by ensuring that the various inserted checks are never bypassed. Each of these capabilities was implemented in the DynamoRIO system with little to no observed overhead.

The commercial potential of the program shepherding system quickly led the authors to found the company Determina, which provided system-wide security policies to be applied to all applications running on a given network. Determina was later acquired by VMware.

A variety of other research groups have since explored several other security applications of dynamic binary modification systems. Various efforts have explored dynamic taint analysis and information flow tracking.

CHAPTER 5

Architectural Exploration

Computer architects have found great utility in dynamic binary modification systems as an enabling technology for fast exploratory studies of novel architectural algorithms or features. These exploratory applications have taken the form of simulation tools, instruction emulation features, and general design-space exploration. Furthermore, dynamic binary modification systems can be used for complete binary translation functionality, to enable a smooth transition from one instruction-set architecture to another, otherwise incompatible architecture. In fact, several corporations have leveraged this precise technology to support industrial strength and widely circulated binary translation (Apple, Dehnert et al. [2003]). The wide variety of opportunities and applications within the computer architecture community will be discussed in this chapter.

5.1 SIMULATION

For the purposes of simulating new or existing architectures, computer architects typically have two options for employing binary modification tools. First, they can build plug-in tools that gather runtime data and feed their custom simulator in real time. Alternatively, they can record runtime data as instruction traces, which can be saved for later use. Figure 5.1 demonstrates the former case of online simulation, while Figure 5.2 depicts the latter case of offline, or trace-driven simulation. Both cases gather the relevant data to drive simulation during an actual run of the guest application. This allows designers to validate that their modifications will not affect the correctness of the guest application, or that the traces they generate represent a valid execution of the user-level code.

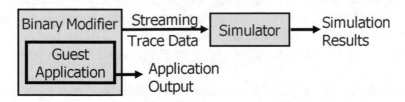

Figure 5.1: Streaming Traces for Online Simulation.

Unlike traditional simulators that can be prohibitively slow, simulators based on dynamic binary modification technology can be on the order of native execution times, allowing more extensive testing than is typically possible. In fact, the robust nature of binary modification systems means that complex applications, such as Oracle database applications, or the user-level components of large

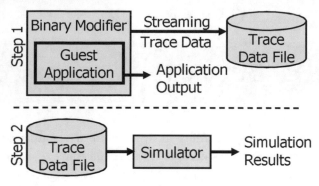

Figure 5.2: Trace Generation for Offline Simulation.

parallel data-mining applications, can be characterized in addition to the small, toy applications, or benchmarks that are commonplace.

5.1.1   TRACE GENERATION

The most straightforward method for leveraging dynamic binary modification to enable simulation is to gather an offline trace that can later be fed into an existing trace-driven simulator, as was depicted in Figure 5.2. The top half of the figure (step 1) involves the use of a dynamic binary modifier to observe execution while recording key events into a trace file. These key events may be as detailed or as high level as necessary, or they can conform to the requirements of the simulator that will be driven by these events. The key events can then be stored in an external file, which can later be used by a simulator, as illustrated in step 2 of the diagram. The simulator can be any arbitrary existing infrastructure or any custom simulator that the user wishes to construct.

While Figure 5.2 depicted a methodology for recording a complete execution trace, it is also possible (and straightforward) to record a representative sample of execution. In fact, tools currently exist to do just this. For instance, the PinPoints system (Patil et al. [2004]) uses Simpoint to locate representative code regions during execution, then uses Pin to record those representative traces for later analysis.

In general, the upside of trace-driven simulation is that it provides a repeatable view of the execution on a guest application. Yet there are two main drawbacks. The first drawback is that it eliminates the possibility of simulating *wrong-path execution* as the outcome of every branch is pre-determined and stored in the trace file, and the result of taking other paths may not be discernible. A second drawback of trace-driven simulation is the need to store the execution trace on disk. These traces can become particularly large, especially for long-running and complex applications, often occupying tens or hundreds of gigabytes of storage, depending on the level of detail recorded. Therefore, streaming this data into a simulator online may be the only practical solution for some applications. The next few sections demonstrate examples of execution-driven simulation that, among other things, will obviate the need for storing traces on disk.

5.1.2 FUNCTIONAL CACHE SIMULATION

While it is possible to simulate a full system, it is also straightforward to provide functional simulation of key components of interest within a computer system, such as the cache and memory system or the branch predictor. Figure 5.3 demonstrates the ease at which a user may develop an execution-driven cache simulator using Pin.

```
                                    ── The DataCache Tool ──
1  VOID MemRef(int tid, ADDRINT addrStart, int size, int type) {
2      for(addr=addrStart; addr<(addrStart+size); addr+=LINE_SIZE)
3          LookupHierarchy( tid, FIRST_LEVEL_CACHE, addr, type);
4  }
5  VOID LookupHierarchy(int tid, int level, ADDRINT addr, int accessType){
6      result = cacheHierarchy[tid][cacheLevel]->Lookup(addr, accessType );
7      if( result == CACHE_MISS ) {
8          if( level == LAST_LEVEL_CACHE ) return;
9          LookupHierarchy(tid, level+1, addr, accessType);
10     }
11 }
12 VOID Instruction(INS ins, VOID *v)
13 {
14     if( INS_IsMemoryRead(ins) )
15         INS_InsertCall(ins, IPOINT_BEFORE, (AFUNPTR) MemRef, IARG_THREAD_ID, IARG_MEMORYREAD_EA,
16                        IARG_MEMORYREAD_SIZE, IARG_UINT32, ACCESS_TYPE_LOAD, IARG_END);
17     if( INS_IsMemoryWrite(ins) )
18         INS_InsertCall(ins, IPOINT_BEFORE, (AFUNPTR) MemRef, IARG_THREAD_ID, IARG_MEMORYWRITE_EA,
19                        IARG_MEMORYWRITE_SIZE, IARG_UINT32, ACCESS_TYPE_STORE, IARG_END);
20 }
21 int main() {
22     PIN_Init();
23     INS_AddInstrumentationFunction(Instruction, 0);
24     PIN_StartProgram();
25     return 0;
26 }
```

Figure 5.3: This figure depicts a snippet of a data cache simulator Pintool, similar to the dcache.cpp tool distributed with Pin.

As Figure 5.3 indicates, cache simulation only requires the dynamic binary modifier to instrument the memory operations present in the guest application. Lines 14 and 17 determine whether a given instruction accesses memory, and if so, whether it is a read or write. This (and only this) information is then forwarded on to the cache simulation engine that runs alongside the guest application. Note that the only memory accesses observed are those accesses by the guest application, and neither the binary modifier itself nor the cache simulator will affect the order or location of the accesses. The cache simulator can then emulate the behavior of the proposed cache hierarchy, determine whether a given read or write would have resulted in a cache miss, and update the simulated cache state accordingly.

A more extensive cache simulator called CMP$im is loosely based on the example above and is presented by Jaleel et al. [2008]. CMP$im simulates multicore caches as well as single core caches.

5.1.3   FUNCTIONAL BRANCH PREDICTION SIMULATION

A similar tactic can be used to simulate the behavior of new branch prediction hardware. The dynamic binary modifier can record branch addresses, targets, and outcomes while running a guest application, and this information can drive a branch prediction simulator. Figure 5.4 illustrates such a simulator.

```
                                    The BrPred Tool
1  BranchPredictor myBPU;
2
3  VOID ProcessBranch(ADDRINT PC, ADDRINT targetPC, bool BrTaken) {
4    BP_Info pred = myBPU.GetPrediction( PC );
5    if( pred.Taken != BrTaken ) {
6      // Direction Mispredicted
7    }
8    if( pred.predTarget != targetPC ) {
9      // Target Mispredicted
10   }
11   myBPU.Update( PC, BrTaken, targetPC);
12 }
13 VOID InstrumentBranches(INS ins, VOID *v)
14 {
15     if( INS_IsDirectBranchOrCall(ins) || INS_HasFallThrough(ins) )
16         INS_InsertCall(ins, IPOINT_BEFORE, (AFUNPTR) ProcessBranch, ADDRINT, INS_Address(ins),
17                     IARG_UINT32, INS_DirectBranchOrCallTargetAddress(ins), IARG_BRANCH_TAKEN, IARG_END);
18 }
19 int main() {
20     PIN_Init();
21     INS_AddInstrumentationFunction(InstrumentBranches, 0);
22     PIN_StartProgram();
23     return 0;
24 }
```

Figure 5.4: Branch Prediction Simulator.

As Figure 5.4 indicates, the dynamic binary modifier focuses only on the branches present in the guest application. It then streams the branch address, target address, and outcome to the branch prediction simulator. The simulator can then emulate the new prediction policy, and it can record the hit/miss rate of the new design. Again, since only branches from the guest application are analyzed, there is no risk of polluting the results with information from the binary modification engine or the simulator itself. And just like the previous simulator, the branch predictor simulation can be performed both online or offline.

5.1.4   TIMING SIMULATION

Finally, a series of small targeted simulators can be combined into a larger functional or even timing simulator. The dynamic binary modifier has the ability to stream any user-level information about the running application to a simulator, so this information can then drive any arbitrary simulator. The user would simply need to know the pertinent information that is necessary to calculate performance and the timing overheads of each performance characteristic.

5.2   EMULATION

Our previous discussions of dynamic binary modification as a driver for architectural simulation emphasized *observing execution* and demonstrated how the observations could then drive various simulators. Another important application of dynamic binary modification is its utility for modifying applications and allowing a user to emulate entirely new functionalities in lieu of the existing functionality that would have occurred otherwise on the underlying machine.

Figure 5.5 shows an example application of emulation. Let's say that a user wishes to replace the normal operation that occurs when loading data from memory with a new implementation. That user could essentially *replace* all instances where data is loaded with a custom implementation. In Figure 5.5, that custom implementation simply augments the load with a logging functionality that also prints out what data was loaded. Yet, the same principle can be applied to replace a load with an entirely new implementation that, for instance, loads all data from a new area of memory.

```
───────────────────────────── The EmuLoad Tool ─────────────────────────────
1  // Moves from memory to register for every executed instruction
2  ADDRINT DoLoad(REG reg, ADDRINT * addr)
3  {
4      cout << "Emulate loading from addr " << addr << " to " << REG_StringShort(reg) << endl;
5      ADDRINT value;
6      PIN_SafeCopy(&value, addr, sizeof(ADDRINT));
7      return value;
8  }
9
10 // This instrumentation routine is invoked once per static instruction
11 VOID EmulateLoad(INS ins, VOID* v)
12 {
13     if (INS_Opcode(ins) == XED_ICLASS_MOV && INS_IsMemoryRead(ins) &&
14         INS_OperandIsReg(ins, 0) && INS_OperandIsMemory(ins, 1)) {
15         // op0 <- *op1
16         INS_InsertCall(ins, IPOINT_BEFORE, AFUNPTR(DoLoad), IARG_UINT32, REG(INS_OperandReg(ins, 0)),
17                         IARG_MEMORYREAD_EA, IARG_RETURN_REGS, INS_OperandReg(ins, 0), IARG_END);
18
19         INS_Delete(ins);
20     }
21 }
22
23 int main(int argc, char * argv[])
24 {
25     PIN_Init(argc,argv);
26     INS_AddInstrumentFunction(EmulateLoad, 0);
27     PIN_StartProgram();
28     return 0;
29 }
```

Figure 5.5: Emulating Loads.

Line 2 of Figure 5.5 begins the new implementation of the load operation, which is inserted as a function call in the guest application. Meanwhile, Line 19 deletes the old implementation of load. All other instructions in the guest application are left untouched.

5.2.1   SUPPORTING NEW INSTRUCTIONS

New processors often come with extensions to the instruction set architecture. This presents a series of challenges to both ISA designers and software vendors. First, ISA designers need an effective way to measure the utility of any new instructions they propose, prior to committing to the actual hardware changes. For instance, it is often helpful to design and optimize the compiler algorithms that would be used to generate the new instructions before committing to building the hardware to support those new instructions. Unfortunately, having a compiler actually generate the new instructions in a binary means that binary cannot be executed until the hardware is available. Otherwise, any attempts to run the program will result in an illegal instruction fault. This *catch 22* situation can be resolved using a dynamic binary modifier, which can recognize and emulate the behavior of any new instructions in a binary that are unsupported by the underlying hardware. Meanwhile, the dynamic binary modification engine can measure the dynamic instruction count and other metrics of the new instructions, providing valuable feedback to the ISA architecture and compiler teams.

Once new instructions have been approved and introduced to the ISA, a second challenge is faced by software vendors, who must now decide whether to include those instructions in their shipped applications. Including the instructions can significantly improve performance on systems that have hardware support for those instructions. Yet, illegal instruction faults will occur on systems that do not support the instructions. Dynamic binary modification provides an elegant solution to these challenges. First, it can provide *backward compatibility*, allowing application binaries that contain new instructions to execute on older hardware that does not support those instructions. Meanwhile, it also provides an opportunity for *forward compatibility* by introducing new instructions to existing binaries on-the-fly, improving the performance of applications running on newer hardware that features extensions to the ISA.

5.2.2   MASKING HARDWARE FLAWS

Many may recall the ordeal in 1994 when a flaw was discovered in the floating-point divide unit of the Pentium processor. This flaw eventually led to a recall of the processor at a cost of several hundred million dollars. A modern day stop-gap solution to such a hardware flaw could be to emulate the correct behavior of the instruction in a software-based virtualization layer, rather than directly executing certain instructions on flawed hardware.

Figure 5.6 demonstrates a simple Pintool that would perform the task of emulating two types of divide instructions that may appear in a guest program. The first case, which is isolated on line 26, finds divide instructions where the operands are both stored in registers. The second case, isolated on line 32, finds divide instructions where one operand is stored in memory. The two replacement routines are found on lines 2 and 10, and each routine emulates the functionality of

The EmuDiv Tool

```
1  // Analysis routines to emulate divide
2  VOID EmulateIntDivide(ADDRINT * pGdx, ADDRINT * pGax, ADDRINT divisor, CONTEXT * ctxt, THREADID tid)
3  {
4      UINT64 dividend = *pGdx;
5      dividend <<= 32;
6      dividend += *pGax;
7      *pGax = dividend / divisor;
8      *pGdx = dividend % divisor;
9  }
10 VOID EmulateMemDivide(ADDRINT * pGdx, ADDRINT * pGax, ADDRINT * pDivisor, unsigned int opSize,
11                       CONTEXT * ctxt, THREADID tid)
12 {
13     ADDRINT divisor = 0;
14     PIN_SafeCopy(&divisor, pDivisor, opSize);
15
16     UINT64 dividend = *pGdx;
17     dividend <<= 32;
18     dividend += *pGax;
19     *pGax = dividend / divisor;
20     *pGdx = dividend % divisor;
21 }
22
23 // Instrumentation routine to replace divide instruction
24 VOID InstrumentDivide(INS ins, VOID* v)
25 {
26     if ((INS_Mnemonic(ins) == "DIV") && (INS_OperandIsReg(ins, 0))) {
27         INS_InsertCall(ins, IPOINT_BEFORE, AFUNPTR(EmulateIntDivide), IARG_REG_REFERENCE, REG_GDX,
28                          IARG_REG_REFERENCE, REG_GAX, IARG_REG_VALUE, REG(INS_OperandReg(ins, 0)),
29                          IARG_CONTEXT, IARG_THREAD_ID, IARG_END);
30         INS_Delete(ins);
31     }
32     if ((INS_Mnemonic(ins) == "DIV") && (!INS_OperandIsReg(ins, 0))) {
33         INS_InsertCall(ins, IPOINT_BEFORE, AFUNPTR(EmulateMemDivide), IARG_REG_REFERENCE, REG_GDX,
34                        IARG_REG_REFERENCE, REG_GAX, IARG_MEMORYREAD_EA, IARG_MEMORYREAD_SIZE,
35                        IARG_CONTEXT, IARG_THREAD_ID, IARG_END);
36         INS_Delete(ins);
37     }
38 }
39
40 int main(int argc, char * argv[])
41 {
42     PIN_Init(argc, argv);
43     INS_AddInstrumentFunction(InstrumentDivide, 0);
44     PIN_StartProgram();    // Never returns
45     return 0;
46 }
```

Figure 5.6: Emulating Divide

the respective divide operation. The ability to easily locate and replace arbitrary instructions from a guest application is a powerful application of dynamic binary modification systems as it enables software solutions to hardware problems.

5.3   BINARY TRANSLATION

The previous section discussed situations where a user wished to replace a small number of key instructions in a shipped binary as it runs. Yet this simple idea can be extended to provide complete binary translation capabilities. An entire binary, designed to be run on one ISA, can be run on an entirely different and incompatible ISA by replacing every instruction encountered with a corresponding instruction or set of instructions. In fact, the translation caching capabilities of most dynamic binary modification systems will result in order of magnitude performance improvements over pure interpretation. This is because sequences of instructions would be translated as a group, and those translated instructions would then be reused throughout the execution time of the guest program.

While most of the widespread binary translation systems of the past, such as Transmeta's Code Morphing Software (Dehnert et al. [2003]), Transitive's Rosetta Software (Apple), or QEMU (Bellard [2005]) have operated below the operating system at the system virtualization level, in principle, there is nothing that prevents a dynamic binary modifier from providing program-at-a-time binary translation capabilities. This approach would require emulation of system libraries and kernel functions, but this has been successfully implemented in products such as Digital's FX!32 (Hookway and Herdeg [1997]).

5.4 DESIGN-SPACE EXPLORATION

Throughout this chapter, we have explored a series of applications of dynamic binary modification technologies within the domain of computer architecture research. The central theme of all of the applications is that the dynamic binary modification system provided for rapid prototyping and design-space exploration. The design process is expedited in two dimensions. First, the high level of abstraction provided by the dynamic binary modifier allows for quick solutions to be designed and built. Meanwhile, the order of magnitude performance improvements over many standard simulators means that designers can explore more dimensions and execute more realistic and longer guest applications than was traditionally possible.

CHAPTER 6

Advanced System Internals

Most users of dynamic binary modification systems do not need to worry about the internal structure and implementation of binary modification engine. In fact, one can argue that the need for understanding of the system itself signifies a weakness of the user-exposed API. Therefore, the content of this chapter is primarily geared toward the researchers and developers of dynamic binary modifiers themselves, while it may also be a useful resource for users who are doing particularly novel or atypical instrumentation tasks.

6.1 MODES OF EXECUTION

There are two fundamental approaches to implementing a dynamic binary modification system - one that is faster but more restrictive, and one that is slower but more robust. In this section, we discuss the main differences between an engine that modifies code in place (probe-mode) versus one that forms a modified copy on demand (JIT-mode).

6.1.1 MODIFIED COPY ON DEMAND

In the most common execution mode, a JIT compiler modifies and recompiles small chunks of binary instructions immediately prior to executing them. Those modified instructions are stored in a software code cache where they are executed in lieu of the original application instructions. Software code caches allow the code regions to be modified once and then reused for the remainder of program execution, helping to amortize the costs of compilation. Most tools use the JIT-based instrumentation since it comes with fewer limitations regarding how and where you can instrument the code.

6.1.2 MODIFICATION IN PLACE

In a probe-based execution mode, the original binary is modified in place (Buck and Hollingsworth [2000], Hunt and Brubacher [1999]). The system overwrites original application instructions with jumps (called probes) to dynamically-generated routines. This code can invoke analysis routines or a replacement routine. Probe-based approaches have near zero execution and memory overhead, but they have a more restrictive API, mainly limiting tools to interposing wrapper routines for global functions while the JIT-based approach allows fine-grained modification down to the instruction level. The reason for the restrictions in probe-mode is obvious in retrospect. On x86, instructions vary in length, and, therefore, not all instructions can be cleanly overwritten by jump instructions. If a branch were to overwrite multiple instructions, then problems would arise if another jump

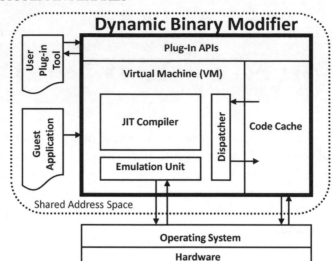

Figure 6.1: Internal organization of a JIT-based dynamic binary modification system. Three programs run in the same address space (the guest application, the plug-in, and the modification engine itself). Within the modification engine, there is a JIT compiler that creates a modified copy of every guest application instruction, a code cache for storing previously translated code, and an emulation unit for maintaining control at system call points.

instruction were to target the second overwritten instructions, as that address would now contain the second half of the new jump instruction.

6.2   A SHARED ADDRESS SPACE

The dynamic binary modifier itself runs in the same address space as both the guest application and the plug-in functionality that the user wishes to add to the guest application, as shown in Figure 6.1. This means that special care must be taken to ensure correctness, isolation, and transparency since the guest application was not designed to share an address space with other applications. If the binary modifier overwrites memory or registers used by the application, then the application could behave differently or incorrectly. Meanwhile, the binary modifier itself requires registers and memory to execute, and, therefore, it must internally manage the process of *context switching* between itself and the guest application. Most systems will attempt to optimize this process, as saving and restoring all state at each of the frequent context switches would be prohibitively expensive. Achieving the various goals (control, isolation, transparency) in a robust manner is nontrivial. The next section describes how these systems acquire control, while later sections discuss isolation and transparency.

6.3   ACQUIRING CONTROL

The dynamic binary modifier will inject itself into the address space of the guest application, much like a debugger, either at the start of program execution, or in some cases, when the user requests it. The former situation is typically called *injection* while the latter situation is called *attach* and *detach*.

Injection is the procedure for loading the dynamic binary modifier into the address space of an application, allowing the system to gain control of execution and then modify and execute the application. It is desirable to perform injection as early as possible so that tools can observe the execution of all guest application instructions. The manner in which injection is achieved varies by modification engine and by operating system.

On Linux, for example, programs are invoked by a `fork` system call, which creates a process, followed by an `exec` system call, which loads a program and starts execution. A process always starts with a single thread of control. When a program executes a clone system call, the kernel creates another thread (called a kernel thread), and it appears as though both the original thread and the newly created thread return from the clone system call. Threads typically end by executing an `exit` system call, which can kill one or all threads.

The Pin dynamic binary instrumentation system intercepts all system calls. When it observes an application `clone` system call, it does the clone on behalf of the application, and it allows the parent thread to resume execution in the code cache. Pin increments the number of threads and allocates a thread-local data structure for the new thread. Pin uses a callback to notify the Pintool when new threads are created. Finally, the new thread resumes executing in the code cache. Pin also intercepts `exit` system calls. It notifies the user's plug-in tool that a thread is exiting and adds the system thread ID to the pending dead list. If this is the only thread or the exit requests to kill all threads, then it also notifies the user's tool that the program is exiting. Finally, it executes the exit system call on behalf of the thread, destroying the kernel thread.

Other systems, such as DynamoRIO, gain control using the `LD_PRELOAD` environment variable on Linux, which informs the dynamic loader to load a shared library into the address space. This is a clean and simple approach that works with dynamically-linked binaries, though not with statically-linked binaries. In addition, some instructions execute before the `LD_PRELOAD` takes effect, and, therefore, it doesn't provide the full coverage that some applications require.

Finally, some systems, such as Pin, support attaching to or detaching from an already running process. The internal methodology is similar to that used by a debugger to attach or detach from running processes.

The Windows operating system presents additional challenges for acquiring control of a guest application. For instance, Pin creates the guest application process in a suspended state and attaches to it using the Win32 debugger APIs (Skaletsky et al. [2010]). The debugger waits for kernel32.dll to complete its initialization routines, and then detaches from the debugged process. Pin saves the application's context and changes the instruction pointer to a small boot routine that is copied to target process memory. The boot routine loads pinvm.dll and calls its main procedure. Pinvm.dll initializes itself and loads the tool DLL. Using this method, tools miss the opportunity to instrument the initialization of the ntdll.dll, kernel32.dll, and kernelbase.dll system libraries, but see every executed instruction of the program binary and DLLs.

DynamoRIO uses a Windows registry key that causes user32.dll to load DLLs listed in the key (Bruening et al. [2001]). This approach works well for programs that load user32.dll. For those programs that do not, they support a backup method which acquires control at the entry point of the application.

6.4   MAINTAINING CONTROL: JIT COMPILATION

Once the dynamic binary modifier has acquired control over the guest application, it maintains control by generating a modified copy of the next few application instructions, and executing those instructions in lieu of the original instructions. The mechanism for doing so is very similar to just-in-time code generation used by Java. However, in a dynamic binary modification system, the input language is binary code, rather than an intermediate format such as Java bytecode. The system decodes the binary instructions to locate branches that will need to be modified while copying all other instructions without changes unless requested by the plug-in tool. All of the branch instructions must be modified to reflect the new location of the modified code, and the fact that the compiler will need to be invoked for any branch targets that have not yet been generated.

Modifications to the guest application are performed on demand as the application executes. Therefore, the run time is spent alternating between (a) generating modified application code and inserting it in the code cache, (b) executing modified application code out of the code cache, (c) transitioning between the two modes, or (d) performing maintenance tasks, such as removing stale code from the code cache or updating the corresponding directories.

6.5   STORING MODIFIED CODE: THE CODE CACHE

In working to reduce overall system overhead, a significant observation is that the single largest performance improvement results from the use of code caches. Software-managed code caches serve the role of storing the modified application code to enable reuse. They improve the overall system performance by amortizing the cost of expensive transformations over the entire program execution time.

The code cache consists of a directory that contains a mapping from original to translated instructions, a set of data structures that keep track of any patched branches in the code cache, and a code area that contains instrumented, translated code as well as auxiliary code to maintain control of execution. The maintenance of the various code cache structures is critical for both correctness and performance (Hazelwood and Smith [2006]). Figure 6.2 shows the space occupied by the translated code, auxiliary code for maintaining control, and the directory data structure needed to track the code cache contents.

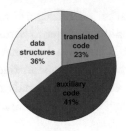

Figure 6.2: Memory distribution of the contents of the software code cache. Contents include translated code, auxiliary code, and data structures.

6.5.1   FORMING TRACES OF MODIFIED CODE

On most systems, the guest application is recompiled one *trace* at a time. A trace is a sequence of extended basic blocks that either reflect the dynamic program path, or the physical adjacency of the original code in memory. In the literature, traces may also be termed *superblocks*. Some systems, such as Valgrind, will recompile traces by raising the binary code into an intermediate format before lowering the instructions back down to binary. At the other extreme, Pin will perform the minimum decoding necessary to locate branches and instructions of interest. And finally, DynamoRIO takes a hybrid approach that allows the user to determine the level of abstraction they wish to manipulate, whether it is a full conversion to intermediate format or somewhere in between.

Stitching together basic blocks into traces is intended to improve the spatial locality of the resulting code in memory, and to enable a series of optimizations that can be applied to the larger code regions. *Trace selection* refers to the act of determining which basic blocks should be included in a trace, and the goal is to capture hot paths – the most frequently-executed sequences of basic blocks in the application.

Duesterwald and Bala [2000] studied the balance between accurate runtime profiling and prediction delays when choosing a trace. They saw that some accuracy could be sacrificed when selecting traces if the trace could be generated sooner because the benefits of forming traces would be gained quicker and leveraged longer. They proposed techniques to reduce the profiling overhead by limiting profiling counters to so-called *trace heads* – the targets of backward branches and the targets of exits from other traces. When those profiling counters reached a certain threshold, they simply formed a trace by following the path of execution during that Nth execution of the instructions. Their trace selection algorithm was termed the Next-Executing Tail (NET) algorithm, and they demonstrated empirically that it outperformed path-profiling based trace selection. This algorithm became the default implementation on the Dynamo dynamic optimizer from Hewlett-Packard, and it remains the algorithm in use by DynamoRIO (a descendent of Dynamo).

Since then, several additional researchers have explored trace selection in dynamic binary modification systems. Berndl and Hendren [2003] and Joshi et al. [2004] explored ways to reduce the overhead of path profiling to make it feasible in a dynamic optimizer. Hiniker et al. [2005]

proposed an alternative to the Next-Executing Tail algorithm called LEI (Last-Executed Iteration), which uses a history buffer during profiling in order to better capture entire loops in a single trace. Meanwhile, Davis and Hazelwood [2011] proposed another alternative to the NET trace selection algorithm (called NETPlus) to better capture loops in traces, which uses a static search to find loop bodies, rather than incurring the persistent overhead of maintaining a history buffer as in the LEI algorithm. Finally, Merrill and Hazelwood [2008] explored the idea of applying trace-selection algorithms to method-based JVMs.

Since the traces stored in the code cache are often formed when the first block in the sequence executes, all subsequent blocks may be speculative. A memory-performance trade-off exists when considering the number of basic blocks to include in a trace. The speculative nature of traces means that various basic blocks that never execute may be compiled and included in a trace. However, when speculation is correct, there are fewer context switches into the VM and the compilation time is well spent. Guha et al. [2010b] explored the delicate balance between trace selection algorithms and memory footprints in the context of embedded systems and other memory-constrained devices, but the same issues persist in general-purpose machines as well.

All of the off-trace paths in the code cache require *exit stubs* (or trampolines) to maintain control and return execution to the VM for on-the-fly compilation of the target. Interestingly, limiting traces to a single basic block results in many more exit stubs (and hence a higher memory footprint) than allowing multi-block traces. This is because a single basic block may end in a conditional branch, which has two potential targets and thus requires two exit stubs. However, forming a contiguous trace would only require one exit stub since the fall through path would remain on trace, obviating the need for us to catch the otherwise off-trace path (Guha et al. [2007]).

When control is transferred from the cached application code, through an exit stub, and back to the binary modifier, then a *context switch* is required to save the register state of the suspended application and set up the register state of the binary modifier. This is an expensive operation, and, therefore, the binary modifier will *link* together traces that execute in succession. This is accomplished by directly patching the cached branch that used to point to an exit stub to now point to the target trace in the cache, obviating the need for a context switch. This act provides a significant speedup and is straightforward for *direct branches* whose target never changes at runtime. Indirect branches, whose target is a register or memory location that changes at runtime, are quite challenging to handle, however, and much of the slowdown of dynamic binary modifiers result from the liberal use of indirect branches in the guest binary. For instance, the `perlbmk` benchmark from SPECint2006 has one of the highest slowdowns on a dynamic binary modifier, and not coincidentally, it also has the largest number of indirect branches encountered at runtime. Hiser et al. [2007] investigated the impact of indirect branch handling in dynamic binary modifiers, while Dhanasekaran and Hazelwood [2011] proposed new mechanisms for improving indirect branch handling in Pin.

Trace linking within the code cache can be performed on-demand (the first time the path is traversed) or proactively (as soon as the source and target are inserted into the code cache). Early (proactive) linking of translated code may create links that are never traversed, and allocate data structures for recording such links. Late (on-demand or lazy) linking does not allocate unnecessary data structures but requires more functionality in the exit stubs, and it also results in more context switches. The three dynamic binary modification systems discussed in throughout the text each take different approaches to linking. Pin links traces proactively; DynamoRIO links traces lazily; Valgrind does not link traces at all, and instead provides a fast context switch which is incurred on every transition between cached traces.

A downside of trace linking is that it requires special care when evicting modified code from the code cache. For instance, if a trace is invalidated, the system also needs to locate all other traces that have patched branches to jump directly to that target trace, and it must restore those branches back to the appropriate exit stubs. Otherwise, control may jump to an invalid location in memory.

6.5.2   CODE CACHE EVICTION AND REPLACEMENT

Many dynamic binary modifiers, such as DynamoRIO and Strata (Scott et al. [2003]), allow their software code caches to grow without bound, as their design highlights performance at the expense of memory footprint. However, when scaling to large applications, unbounded code caches are beneficial neither from the performance nor the memory footprint perspective. In fact, even the simple, single-threaded application perlbench from SPEC2006 exhibits a speedup when the cache is bounded. Similar trends can be observed on larger enterprise applications that exhibit little code reuse. Intuitively, those applications have no need to maintain the translated code sequences that are a part of an initialization sequence, and meanwhile, those sequences take up valuable memory and hardware cache resources.

Aside from bounding the size of the code cache, there are numerous reasons that translated code must be removed from the code cache. Dynamically unloaded code will need to be invalidated as other application code may be loaded at the same address. Users may request invalidation of cached code to remove instrumentation or to re-optimize regions of code. Finally, self-modifying code must also trigger invalidation and regeneration of the corresponding cached code.

The most straightforward code cache replacement algorithm is to flush the entire code cache when the free space is depleted. This approach is superior to other intuitive algorithms, such as LRU replacement, given that the stored traces vary considerably in size, and, therefore, individual trace deletions will lead to fragmentation.

However, full cache flushes are not possible for systems that support multithreaded guest applications where the various threads share a single underlying code cache and since it is expensive to determine whether any of the inactive threads are currently executing the code that is to be deleted. The issue of multithreading and code caches is discussed in more detail in Section 6.7.

Meanwhile, there are many granularities of code cache eviction algorithms that span the spectrum from full cache flushes to individual trace deletions (Guha et al. [2010a], Hazelwood and Smith [2004]). Traces can be grouped into large, fixed-sized *cache blocks*, and entire cache blocks can be flushed at once. This approach reduces the number of cache misses compared to a full flush, without

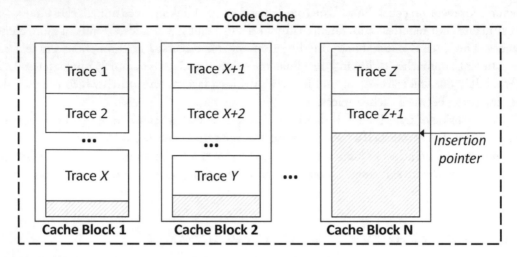

Figure 6.3: Organization of a software code cache that supports medium-grained evictions. The traces are grouped into larger, fixed-sized cache blocks, which can be deleted as a whole, avoiding fragmentation.

suffering the fragmentation issues that hinder fine-grained evictions. Figure 6.3 presents a code cache organization that supports medium-grained evictions.

For the interested reader, a much more detailed discussion of code cache replacement issues and policies is available in the Ph.D. thesis by Hazelwood [2004] or in the papers by Hazelwood and Smith [2006] or Bruening and Amarasinghe [2005].

6.5.3   CODE CACHE INTROSPECTION

Most frameworks have gone to great lengths to mask the presence of a code cache from the user by converting cached instruction addresses to their corresponding addresses in the original application before presenting the addresses to the user. However, providing researchers access to the contents of the code cache enables many powerful opportunities. A user can manipulate the code cache contents to investigate run-time optimizations or security policies; they can instrument and compare applications across several architectures; they can even investigate the code cache implementation itself and develop and compare custom code cache replacement policies.

Several dynamic binary modification systems provide a glimpse into, or even permit control of, the code cache and its operation. For instance, DynamoRIO provides logging capabilities that allow a user to observe the code cache contents during program execution. In addition, DynamoRIO users can modify the code cache algorithms in place since source code is available for the internal engine. Meanwhile, the Pin system provides a wide API that provides four categories of code cache introspection. *Callbacks* will notify the Pin user when various events occur within the code cache, such as trace insertion, deletion, or execution. The callbacks give the user tool an opportunity to

```
void InsertSmcCheck () {
  traceAddr = (VOID *)TRACE_Address(trace);
  traceSize = TRACE_Size(trace);
  traceCopyAddr = malloc(traceSize);
  if (traceCopyAddr != 0) {
    memcpy(traceCopyAddr, traceAddr, traceSize);

    // Insert DoSmcCheck call before every trace
    TRACE_InsertCall(trace, IPOINT_BEFORE, (AFUNPTR)DoSmcCheck, IARG_PTR, traceAddr,
        IARG_PTR, traceCopyAddr, IARG_UINT32, traceSize, IARG_CONTEXT, IARG_END);
  }
}
VOID DoSmcCheck(VOID * traceAddr, VOID * traceCopyAddr, USIZE traceSize, CONTEXT * ctxP) {
  if (memcmp(traceAddr, traceCopyAddr, traceSize) != 0) {
    smcCount++;
    free(traceCopyAddr);
    CODECACHE_InvalidateTrace((ADDRINT)traceAddr);
    PIN_ExecuteAt(ctxP);
  }
}
void main (int argc, char **argv) {
  PIN_Init(argc, argv);
  TRACE_AddInstrumentFunction(InsertSmcCheck,0);
  PIN_StartProgram(); // Never returns
}
```

Figure 6.4: A code cache introspection tool that detects and handles self-modifying code.

perform custom *actions*, such invalidating a single trace, or even flushing the entire code cache. The *lookup* API provide access to Pin's internal data structures that keep track of the code cache's contents, and finally, the *statistics* API gives the user access to aggregated data about the various actions that have taken place at any given point in time. More details about the potential applications of cache introspection is discussed by Hazelwood and Cohn [2006].

6.5.4 HANDLING SELF-MODIFYING CODE

Self-modifying code is a challenge to handle efficiently in a dynamic binary modifier (Bruening and Amarasinghe [2005], Dehnert et al. [2003]). The problem occurs when an application executes some code, modifies it, and then executes the new code at the same address. After the first execution, a dynamic binary modifier will save a copy of the code in its code cache. When the modified code is executed, the system translator must detect that the code it has in its cache is no longer valid. Without any detection, it will continue to execute the old code, and the program will eventually fail.

Several mechanisms have been proposed to detect self-modifying code such as write-protecting code pages, checking store addresses, and inserting extra code to check that instruction memory has not changed. A straightforward (but admittedly inefficient) solution can be written in 15 lines of code using the code cache introspection API described in the previous section. Figure 6.4 shows a simple self-modifying code handler. The function `InsertSmcCheck` is the instrumentation function which is passed a list of instructions in the trace. While an instrumentation function typically inserts calls to count basic block executions or record the effective address for a memory reference, this particular function makes a copy of the original instructions in the trace and inserts a call to DoSmcCheck, passing it the address in memory of the instructions and the saved copy. When the trace is executed, it calls DoSmcCheck. This function compares the current contents of the instruction memory against the saved copy. If it has changed, it invalidates the cached copy of the trace and uses PIN_ExecuteAt (Pan et al. [2005]) to re-invoke the trace. Note that this example does not handle a trace that overwrites its own code (after the check) or multithreaded guest applications. It is also possible to use page protection by instrumenting memory management system calls. Mechanisms that watch store addresses can be implemented by instrumenting memory store instructions.

6.6   THE EMULATOR

System calls are triggered by instructions that transition from executing an application in user mode to executing the kernel in system mode. They are used to request system services like file I/O and process creation. Since a dynamic binary modifier operates at the process level, it has full control of everything that executes in user mode, but it loses control in kernel mode. It must manage the execution of system calls to ensure to regain control when the system resumes execution in user mode. There are three steps to managing system calls. First, the binary modifier must detect when the application is about to execute a system call, and instead redirect control to the compilation engine. Second, it must be able to execute the system call on behalf of the application. Third, it must be able to regain control after an interrupted system call; when a system call is interrupted, the kernel may cause the application to continue execution elsewhere. The system must intercept execution after an interrupted system call to direct control back to its code cache. Meanwhile, at all points during the execution of a system call, the binary modifier must be able to construct a precise state reflecting the application register values without the effects of instrumentation.

The dynamic binary modifier needs to detect and intercept system call instructions at a point where it can capture the arguments and system call number, and then transfer control to the compilation engine. Since all user-mode instructions are inspected and placed into the code cache, this becomes the logical place to intercept system calls as the system can easily detect the instructions that transfer control from user mode to the kernel.

On Linux, system calls are straightforward to detect. The int 0x80, sysenter, and syscall instructions will transfer control to the kernel after placing the system call number into the eax register.

On Windows, the sysenter or int 0x2e instructions are used to transfer control from user mode to the kernel on 32-bit systems, and the syscall instruction is used for 64-bit processes. When a 32-bit Windows process runs on 64-bit Windows, it uses the jmp far instruction to enter the kernel through the WOW64 layer. This instruction is located at a well-known address that is stored in the Thread Environment Block (TEB) of each thread, so it can be easily recognized. While it is also possible to intercept system calls using the Win32 API, most dynamic binary modifiers will not operate at this level since the Win32 API can be (and often is) easily bypassed, causing the system to miss events and lose control of the executing application.

Unix signals also present a challenge to a dynamic binary modifier. Allowing the guest application to simply handle signals natively would present three problems. First, the application's signal handler escapes from the control of the dynamic binary modifier and would therefore be executed without the opportunity for modification. Second, if the application's handler inspects the interrupted register context, it sees the interrupted state of the binary modifier, not the pure application state. Finally, the signal might interrupt the user's plug-in functionality and cause reentrancy problems. For these reasons, the emulator must intercept the application's signal handlers and emulate signal delivery. In Pin, for example, when a signal arrives, Pin catches the signal and places it on a queue of pending signals. Execution then resumes until any tool instrumentation completes and control reaches the end of the next code cache trace. At the end of the trace, control jumps to the Pin signal emulator, which builds an emulated signal context frame using the values of the application's registers. Finally, the Signal Emulator starts translating code at the PC of the application's handler.

6.7   MULTITHREADED PROGRAM SUPPORT

Much of the literature describing the internal architecture and performance of dynamic binary modifiers has focused on executing single-threaded guest applications. This section discusses the specific design decisions necessary for supporting large, multithreaded guest applications. While implementing a working solution for multithreading is straightforward, providing a system that scales in terms of memory and performance is much more intricate.

6.7.1   THREAD-SHARED CODE CACHES

It is fairly straightforward to implement a code cache in such a way that each guest application thread has its own code cache. This approach avoids many of the synchronization and deletion headaches that arise when multiple threads share a cache. For instance, the task of deleting a single trace from the code becomes fairly complicated to do efficiently, since the system must ensure that no other threads are executing or stalled within the candidate trace. In addition, patching a branch within a trace so that it will jump directly to another trace must be done in an atomic manner, such that other threads do not inadvertently execute partially-modified code.

Despite the complexity of thread-shared code caches, researchers have demonstrated (Bruening et al. [2006]) that thread-private code caches do not scale beyond a very small number of threads. Therefore, it is worth the development effort to employ a single, shared code cache across all threads. For the applications where several threads perform similar tasks, the shared cache contains all of the common code, and thus the memory footprint is smaller, and more scalable with the number of threads. Figure 6.5 demonstrates the code expansion that occurs when the SPEC OMP 2001 benchmarks are executed using thread-private code caches and eight application threads. The figure shows that the code can expand well over 500% if we allow each thread to have

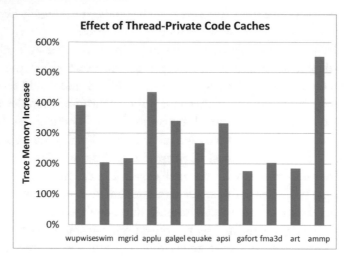

Figure 6.5:   Code expansion resulting from thread-private code caches. The SPEC OMP 2001 bench-marks were run as 8 application threads.

its own private cache. This memory scalability issue justifies the use of a shared code cache across threads.

6.7.2   GENERATIONAL CACHE REPLACEMENT

One fundamental observation is that code cache maintenance becomes much more complex when executing a multithreaded application. For instance, a complete flush of the code cache is no longer possible. The problem is each thread is in one of two states – executing code in the code cache, or generating new code in the compilation engine. In fact, many threads will be stalled in one of these two states. The system may not delete a trace that another thread is currently executing or stalled in, but it is also extremely difficult to determine where each thread is executing without employing very expensive techniques like having every thread log every trace it enters. While an easy alternative is to stall the flush until all threads are executing inside the modification engine code (and thus not executing cached, modified application code), this stall can take an indefinite amount of time, and meanwhile threads are not able to make forward progress.

The standard solution is to employ a generational code cache deletion algorithm, demonstrated in Figure 6.6. The code cache is partitioned into multiple blocks, and each block is marked with a generation number. Initially, all threads execute in generation 1. When a code cache flush needs to occur, we can simply clear the cache lookup table (so that no threads enter the cache blocks for the current generation), advance the generation counter, and create a new cache block tagged with the new cache generation. As each thread enters the modification engine, it will only find the new traces present in the new generation since the old traces will have been removed from the look-up

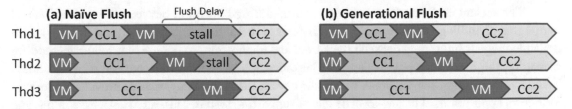

Figure 6.6: Timeline comparing a naïve code cache flush to a thread-safe generational flush. The naïve implementation stalls until all threads return to the VM, while the generational implementation makes forward progress as it waits for threads to return to the VM.

table, and thus the thread will never re-enter code from the old generation. Threads may also move forward with generating new code for the new generation. Finally, when the last thread leaves the old generation, we flush the cache blocks for that generation. This scheme allows threads to continue to generate and execute new code while other threads are potentially stalled and/or in the process of leaving the old generation.

More details about the support necessary for handling multithreaded applications can be found in a paper by Hazelwood et al. [2009].

6.8   WINDOWS EXECUTION SUPPORT

Analyzing the behavior of Windows applications is an important challenge because the guest application source code is not typically available, and the Windows kernel interface cannot be adapted to support observability. The fact that a dynamic binary modifier can instrument unmodified binaries means that it becomes possible to analyze the performance of proprietary Windows applications in realistic scenarios. This section identifies the Windows-specific obstacles for implementing a dynamic binary modifier.

The first challenge involves getting initial control of the process. This is called injection, and it was discussed earlier in Section 6.3. To provide the maximum observability, the system must inject itself into a new process as early as possible. Some systems depend on some basic Windows services, so it is preferable from an implementation perspective to delay injection until various services are initialized. However, late injection provides less observability. Aside from injecting into the parent process, it is important to inject into any child process that gets spawned at run time. An issue on Windows is that processes can be 32-bit, 64-bit, or even a combination of both, so the system needs to be aware of and support both modes. Creating a child process usually requires a sequence of three system calls: `NtCreateProcess`, `NtCreateThread`, and `NtResumeThread`. The dynamic binary modifier can choose any point in that sequence to take control. For instance, Pin alters the child process state immediately before `NtResumeThread` while DynamoRIO alters it during `NtCreateThread`.

The second challenge is maintaining control during transitions between the application and kernel (e.g., handling system calls). The Windows kernel interface is not designed to have an independent agent interposed between the kernel and application, so designing a robust method for managing the transitions is a challenge. It becomes necessary to venture outside of the standard Windows API to get the job done. Below the covers, there are many ways to provide support for Windows applications. The low-level details of supporting system calls, interrupts, exceptions, and system library calls on Windows are detailed in papers by Skaletsky et al. [2010] for Pin, and by Bruening et al. [2001] for DynamoRIO. The PhD thesis by Bruening [2004] also provides additional details on the DynamoRIO implementation.

6.9   MASKING OVERHEAD WITH PARALLELISM

When looking at the overhead of leveraging a dynamic binary modifier to perform a given task, it is important to note that the observed overhead comes from three different sources: (1) the baseline overhead of the native guest application; (2) the overhead added by executing under the control of a dynamic binary modifier, which can cause slowdowns on the order of 1X - 10X depending on the features of the guest application and the binary modifier chosen; and (3) the overhead of the additional functionality injected by the user, such as performing program analysis or running simulations, which can cause slowdowns on the order of 1000X or more. So far, the techniques discussed in this chapter have focused on the second form of overhead incurred by the binary modifier itself. However, an opportunity exists to mask the overhead of the third form of overhead by leveraging the available parallelism available on the underlying hardware.

Normally, dynamic binary modifiers execute one copy of the instrumented application in a serial fashion, alternating between the three sources of overhead discussed above. However, another alternative is to separate the tasks of instrumentation and analysis into separate threads and to routinely fork off the task of analyzing a portion of the guest application. These analyses can then be overlapped in time by leveraging separate processing resources. Depending on the number of available processor cores and memory, and the size of each slice, it is possible to approach the native, uninstrumented execution speed of the application, even for complex analysis tasks or simulations. Meanwhile, it is never possible to execute faster than native since one thread is always the native, uninstrumented application running serially.

Three different sets of researchers have explored this general notion, and the three approaches taken have subtle but important differences. Wallace and Hazelwood [2007] implemented SuperPin, which runs the guest application without instrumentation, then uses `ptrace` to routinely spawn off instrumented `slices` of the application that represents a set of distinct, non-overlapping portions of the execution time. Special care must be taken to ensure that it is safe to execute the two copies of the guest application (one native and one instrumented) without affecting correctness, and this is all handled transparently by SuperPin. As expected, the overall solution came with a large set of design and implementation challenges. Design decisions included the granularity and frequency at which the instrumented slices should be spawned. Less frequent (longer) slices minimized the overhead caused by using `fork`, but it meant that a lower degree of parallelism could be achieved. Furthermore, synchronizing multiple instrumentation slices, handling system calls, and merging results from the various instrumentation slices were but a few of the challenges that had to be tackled during the implementation phase.

While SuperPin tried to be a general solution, the reality was that a major impediment was the dependencies between the various slices. For instance, at any time, a cache simulation will depend on the result of previous cache activity, and some of that activity will be unknown if the instrumentation slices are executed in a pipelined manner. As a result, Moseley et al. [2007] focused on the goal of sampling runtime performance features of a guest application, which is a common use of a dynamic binary modifier. In this case, full coverage of the guest application is not necessary, and, therefore, there are no inter-slice dependencies. They were able to explore the trade-off between accuracy and overhead when using parallelism to hide the sampling overhead.

Finally, Zhao et al. [2010] used DynamoRIO to instrument a guest application, and then manually pipelined the analysis routines to hide their overhead. Again, the final overhead depended on how well balanced the various pipeline stages were, but it was never faster than native performance. This was true for all three bodies of work discussed in this section.

6.10   REMAINING CHALLENGES

Getting a dynamic binary modifier to "work" is actually fairly straightforward. The bulk of the challenge lies in getting the system to "work well" - both in handling all of the various corner cases, and in doing so efficiently. The goal of improving the performance of the overall system will always be a standing challenge, as the chosen solutions for trace selection, cache management, and handling of user-defined analysis routines will always be able to be further refined and optimized. Most of the solutions thus far have focused on optimizing the systems on stock hardware. However, more work should be done to propose new hardware features that will greatly reduce the overhead of the more expensive tasks performed, such as context switching between the various forms of runtime overhead, handling indirect branches, handling self-modifying code, and handling the modified code cache.

CHAPTER 7

Historical Perspectives

The dynamic binary modification systems detailed in this text are by no means the first of their kind (nor are they likely to be the last). The three systems were chosen as the focus of this book because at the time of its writing, they were widely used and readily available.

In the 1990's, several other dynamic binary modifiers were developed, including Shade for SPARC/Solaris (Cmelik and Keppel [1994]), DynInst for a variety of platforms (Buck and Hollingsworth [2000]), Vulcan for x86/Windows (Edwards et al. [2001]), Wiggins/Redstone for Alpha (Deaver et al. [1999]), and Dynamo for HPUX/PA-RISC (Bala et al. [1999]).

Later on, numerous other dynamic binary instrumentation frameworks appeared, including Strata (Scott et al. [2003]), DELI (Desoli et al. [2002]), which is a descendent of Dynamo for the LX architecture, DIOTA for x86/Linux (Maebe et al. [2002]), Mojo for x86/Windows (Chen et al. [2000]), Walkabout for SPARC/Solaris (Cifuentes et al. [2002]), and HDTrans (Sridhar et al. [2006]). In addition, the three focus systems from this book (Pin, DynamoRIO, and Valgrind) appeared during that first decade of 2000.

Other tools served a similar purpose to one of more of the applications of dynamic binary modification, such as simulation or dynamic translation. Hardware simulators or emulators include Embra (Witchel and Rosenblum [1996]) and Simics (Magnusson et al. [2002]). Dynamic binary translators include DAISY (Ebcioğlu and Altman [1997]), Crusoe (Dehnert et al. [2003]), and Rosetta (Apple).

Outside of dynamic binary modification, there are a wide variety of static instrumentation tools dating back several decades. For instance, the ATOM toolkit from Digital (Srivastava and Eustace [1994]) formed the basis for the look-and-feel of the Pin API, and indeed there were several developers in common. Meanwhile, other systems included Etch (Romer et al. [1997]), EEL (Larus and Schnarr [1995]), and Morph (Zhang et al. [1997]).

CHAPTER 8

Summary and Observations

The "compile once, run anywhere" philosophy has imposed a number of stumbling blocks for modern systems and software. As computer architectures have evolved and software has become significantly more complex, the need to completely understand and potentially modify the runtime behavior of modern software has become paramount. Unfortunately, the standard software distribution model hinders this goal as software is often distributed in binary form, with relevant information necessary for analyzing or modifying the application permanently removed. Fortunately, dynamic binary modification has emerged as a means for bypassing the restrictions of binary code, and accomplishing a series of tasks that were never envisioned and perhaps deemed impossible by the initial designers of computer systems and software.

Since their introduction many years ago, dynamic binary modifiers have gained significant popularity due to their robustness, ease-of-use, and widespread utility – which spans several subdisciplines within computer science and engineering. Users have discovered that these systems can be leveraged for fast and detailed program analysis, arbitrary runtime modification, and architectural design space exploration to name a few. As a result, the use of dynamic binary modifiers has become ubiquitous. In fact, in many recent technical conferences focusing on computer systems, roughly two-thirds of the published papers have used some form of dynamic binary modification to explore and evaluate their proposed solutions[1].

Modern dynamic binary modifiers are quite robust, and they can correctly execute and modify nearly every application in use today. This is an impressive feat, given the internal design complexity as well as the complexity of the various tasks that each system supports. For instance, transparently handling complex, but common cases such as observing and modifying dynamically-loaded and shared libraries or dynamically generated code requires a significant engineering effort within the modification engine itself. Meanwhile, all of this complexity is masked from the end user, allowing them to perform complex and invasive tasks with very little effort. A user can get up to speed and build useful tools in a matter of hours. Learning to use the system in the most effective manner takes some additional understanding, however, and the goal of this text was to provide that firm foundation of knowledge for the end user.

The bulk of the researchers interested in dynamic binary modification see it as a useful tool for their own research goals and are therefore focused on the end-user view and applications. However, a growing community of advanced users and system software developers are interested in understanding and/or potentially optimizing the internal design of dynamic binary modification systems

---

[1]This claim is based on informal observations made at MICRO 2009, HPCA 2010, ASPLOS 2010, CGO 2010, and ISCA 2010.

themselves. The book's coverage of the internal structures and algorithms is intended to meet that goal. While many of the individual components have been covered in one form or another throughout the body of literature in place today, the goal of this text was to provide a comprehensive and up-to-date view of the relevant structures and algorithms in place in some of the most commonly used dynamic binary modification systems today.

Above all, the subfield of dynamic binary modification is still evolving. New applications, new challenges, and new internal algorithms are regularly surfacing. One of the luxuries of an electronically-based textbook series is that the text itself can and should evolve as well. As such, I hope to include your own breakthroughs in future editions. Stay tuned!

ADDITIONAL RESOURCES

This book provides detailed coverage of one of many topics covered in the virtual machines textbook by Smith and Nair [2005]. The reader can refer to their book to get an understanding of how dynamic binary modification fits into the bigger virtualization picture.

For the most part, one seminal paper exists for each of the three dynamic binary modification systems covered in this book. For Valgrind, that seminal paper appeared in PLDI 2007 (Nethercote and Seward [2007]), for Pin, it appeared in PLDI 2005 (Luk et al. [2005]), and for DynamoRIO, it appeared in CGO 2003 (Bruening et al. [2003]). Readers interested in one particular system should start by focusing on the seminal work before moving on to the large number of followup papers that have appeared since then. The interested reader should also consider reading the Ph.D. theses of some of the developers of the three systems highlighted in this text. In particular, Bruening [2004] presents an in-depth look at the internal workings of DynamoRIO. Meanwhile, Nethercote [2004] presents the internal workings of Valgrind, as well as a nice historical perspective on similar systems.

Bibliography

Apple. Rosetta. http://www.apple.com/rosetta/. 31, 38, 55

Moshe (Maury) Bach, Mark Charney, Robert Cohn, Elena Demikhovsky, Tevi Devor, Kim Hazel-wood, Aamer Jaleel, Chi-Keung Luk, Gail Lyons, Harish Patil, and Ady Tal. Analyzing parallel programs with Pin. *IEEE Computer*, 43(3):34–41, March 2010. DOI: 10.1109/MC.2010.60 23

Vasanth Bala, Evelyn Duesterwald, and Sanjeev Banerjia. Transparent dynamic optimization. Technical Report HPL-1999-77, Hewlett Packard, June 1999. 27, 55

Vasanth Bala, Evelyn Duesterwald, and Sanjeev Banerjia. Dynamo: a transparent dynamic optimization system. In *Proceedings of the ACM SIGPLAN Conference on Programming Language Design and Implementation*, PLDI '00, pages 1–12, Vancouver, BC, Canada, June 2000. DOI: 10.1145/349299.349303 27

Utpal Banerjee, Brian Bliss, Zhiqiang Ma, and Paul Petersen. A theory of data race detection. In *Proceedings of the 2006 Workshop on Parallel and Distributed Systems: Testing and Debugging*, PADTAD '06, pages 69–78, Portland, ME, USA, July 2006. DOI: 10.1145/1147403.1147416 23

Fabrice Bellard. Qemu: a fast and portable dynamic translator. In *Proceedings of the USENIX Annual Technical Conference*, ATEC '05, pages 41–46, Anaheim, CA, USA, 2005. USENIX Association. 38

Marc Berndl and Laurie Hendren. Dynamic profiling and trace cache generation. In *Proceedings of the 1st Annual IEEE/ACM International Symposium on Code Generation and Optimization*, CGO '03, pages 276–288, San Francisco, CA, USA, March 2003. DOI: 10.1109/CGO.2003.1191552 43

Derek Bruening and Saman Amarasinghe. Maintaining consistency and bounding capacity of software code caches. In *Proceedings of the 3rd Annual IEEE/ACM International Symposium on Code Generation and Optimization*, CGO '05, pages 74–85, San Jose, CA, USA, March 2005. DOI: 10.1109/CGO.2005.19 46, 47

Derek Bruening and Vladimir Kiriansky. Process-shared and persistent code caches. In *Proceedings of the 4th Annual ACM SIGPLAN/SIGOPS International Conference on Virtual Execution Environments*, VEE '08, pages 61–70, Seattle, WA, USA, March 2008. DOI: 10.1145/1346256.1346265 8

Derek Bruening, Evelyn Duesterwald, and Saman Amarasinghe. Design and implementation of a dynamic optimization framework for windows. In *Proceedings of the 4th ACM Workshop on Feedback-Directed and Dynamic Optimization*, FDDO-4, Austin, TX, USA, December 2001. 42, 52

Derek Bruening, Timothy Garnett, and Saman Amarasinghe. An infrastructure for adaptive dynamic optimization. In *Proceedings of the 1st Annual IEEE/ACM International Symposium on Code Generation and Optimization*, CGO '03, pages 265–275, San Francisco, CA, USA, March 2003. DOI: 10.1109/CGO.2003.1191551 58

Derek Bruening, Vladimir Kiriansky, Timothy Garnett, and Sanjeev Banerji. Thread-shared software code caches. In *Proceedings of the 4th Annual IEEE/ACM International Symposium on Code Generation and Optimization*, CGO '06, pages 28–38, New York, NY, USA, March 2006. DOI: 10.1109/CGO.2006.36 49

Derek L. Bruening. *Efficient, Transparent and Comprehensive Runtime Code Manipulation*. PhD thesis, Massachusetts Institute of Technology, Cambridge, MA, USA, September 2004. 52, 58

Bryan Buck and Jeffrey K. Hollingsworth. An api for runtime code patching. *International Journal of High Performance Computing Applications*, 14(4):317–329, November 2000. DOI: 10.1177/109434200001400404 39, 55

Wen-Ke Chen, Sorin Lerner, Ronnie Chaiken, and David Gillies. Mojo: A dynamic optimization system. In *Proceedings of the 4th ACM Workshop on Feedback-Directed and Dynamic Optimization*, FDDO-4, pages 81–90, Austin, TX, USA, December 2000. 55

Cristina Cifuentes, Brian Lewis, and David Ung. Walkabout: A retargetable dynamic binary translation framework. Technical Report SMLI TR-2002-106, Mountain View, CA, USA, 2002. 55

Bob Cmelik and David Keppel. Shade: A fast instruction-set simulator for execution profiling. In *Proceedings of the ACM SIGMETRICS Conference on Measurement and Modeling of Computer Systems*, SIGMETRICS '94, pages 128–137, Nashville, TN, USA, May 1994. ACM. DOI: 10.1145/183018.183032 55

Derek Davis and Kim Hazelwood. Improving region selection through loop completion. In *Proceedings of the ASPLOS Workshop on Runtime Environments/Systems, Layering, and Virtualized Environments*, RESoLVE '11, Newport Beach, CA, USA, March 2011. 44

Dean Deaver, Rick Gorton, and Norm Rubin. Wiggins/redstone: An on-line program specializer. In *IEEE Hot Chips XI*, 1999. 55

James C. Dehnert, Brian K. Grant, John P. Banning, Richard Johnson, Thomas Kistler, Alexander Klaiber, and Jim Mattson. The transmeta code morphing software: Using speculation, recovery, and adaptive retranslation to address real-life challenges. In *Proceedings of the 1st Annual*

IEEE/ACM International Symposium on Code Generation and Optimization, CGO '03, pages 15–24, San Francisco, CA, USA, March 2003. DOI: 10.1109/CGO.2003.1191529 31, 38, 47, 55

Giuseppe Desoli, Nikolay Mateev, Evelyn Duesterwald, Paolo Faraboschi, and Joseph A. Fisher. Deli: A new run-time control point. In *Proceedings of the 35th Annual ACM/IEEE International Symposium on Microarchitecture*, MICRO-35, pages 257–268, Istanbul, Turkey, 2002. DOI: 10.1109/MICRO.2002.1176255 55

Balaji Dhanasekaran and Kim Hazelwood. Improving indirect branch translation in dynamic binary translators. In *Proceedings of the ASPLOS Workshop on Runtime Environments/Systems, Layering, and Virtualized Environments*, RESoLVE '11, Newport Beach, CA, USA, March 2011. 44

Evelyn Duesterwald and Vasanth Bala. Software profiling for hot path prediction: Less is more. In *Proceedings of the 12th International Conference on Architectural Support for Programming Languages and Operating Systems*, ASPLOS '00, pages 202–211, Cambridge, MA, USA, October 2000. DOI: 10.1145/356989.357008 43

Kemal Ebcioğlu and Erik R. Altman. Daisy: dynamic compilation for 100% architectural compatibility. In *Proceedings of the 24th Annual International Symposium on Computer Architecture*, ISCA '97, pages 26–37, Denver, CO, USA, 1997. ACM. DOI: 10.1145/384286.264126 55

Andrew Edwards, Amitabh Srivastava, and Hoi Vo. Vulcan: Binary transformation in a distributed environment. Technical Report MSR-TR-2001-50, Microsoft Research, April 2001. 55

Apala Guha, Kim Hazelwood, and Mary Lou Soffa. Reducing exit stub memory consumption in code caches. In *Proceedings of the International Conference on High-Performance Embedded Architectures and Compilers*, HiPEAC '07, pages 87–101, Ghent, Belgium, January 2007. DOI: 10.1007/978-3-540-69338-3_7 44

Apala Guha, Kim Hazelwood, and Mary Lou Soffa. Balancing memory and performance through selective flushing of software code caches. In *Proceedings of the International Conference on Compilers, Architectures and Synthesis for Embedded Systems*, CASES '10, pages 1–10, Scottsdale, AZ, USA, October 2010a. DOI: 10.1145/1878921.1878923 45

Apala Guha, Kim Hazelwood, and Mary Lou Soffa. Dbt path selection for holistic memory efficiency and performance. In *Proceedings of the 6th ACM SIGPLAN/SIGOPS International Conference on Virtual Execution Environments*, VEE '10, pages 145–156, Pittsburgh, PA, USA, March 2010b. DOI: 10.1145/1837854.1736018 44

Kim Hazelwood. *Code Cache Management in Dynamic Optimization Systems*. PhD thesis, Harvard University, Cambridge, MA, USA, May 2004. 46

Kim Hazelwood and Robert Cohn. A cross-architectural framework for code cache manipulation. In *Proceedings of the 4th Annual IEEE/ACM International Symposium on Code Generation and Optimization*, CGO '06, pages 17–27, New York, NY, USA, March 2006. DOI: 10.1109/CGO.2006.3 47

Kim Hazelwood and Artur Klauser. A dynamic binary instrumentation engine for the arm architecture. In *Proceedings of the International Conference on Compilers, Architectures, and Synthesis for Embedded Systems*, CASES '06, pages 261–270, Seoul, Korea, October 2006. DOI: 10.1145/1176760.1176793 13

Kim Hazelwood and James E. Smith. Exploring code cache eviction granularities in dynamic optimization systems. In *Proceedings of the 2nd Annual IEEE/ACM International Symposium on Code Generation and Optimization*, CGO '04, pages 89–99, Palo Alto, CA, USA, March 2004. DOI: 10.1109/CGO.2004.1281666 45

Kim Hazelwood and Michael D. Smith. Characterizing inter-execution and inter-application optimization persistence. In *Proceedings of the Workshop on Exploring the Trace Space for Dynamic Optimization Techniques*, pages 51–58, San Francisco, CA, USA, June 2003. 8

Kim Hazelwood and Michael D. Smith. Managing bounded code caches in dynamic binary optimization systems. *Transactions on Code Generation and Optimization*, 3(3):263–294, September 2006. DOI: 10.1145/1162690.1162692 42, 46

Kim Hazelwood, Greg Lueck, and Robert Cohn. Scalable support for multithreaded applications on dynamic binary instrumentation systems. In *Proceedings of the ACM International Symposium on Memory Management*, ISMM '09, pages 20–29, Dublin, Ireland, June 2009. DOI: 10.1145/1542431.1542435 51

David J. Hiniker, Kim Hazelwood, and Michael D. Smith. Improving region selection in dynamic optimization systems. In *Proceedings of the 38th Annual International Symposium on Microarchitecture*, MICRO-38, pages 141–154, Barcelona, Spain, November 2005. DOI: 10.1109/MICRO.2005.22 43

Jason D. Hiser, Daniel Williams, Wei Hu, Jack W. Davidson, Jason Mars, and Bruce R. Childers. Evaluating indirect branch handling mechanisms in software dynamic translation systems. In *Proceedings of the 5th Annual IEEE/ACM International Symposium on Code Generation and Optimization*, CGO '07, pages 61–73, San Jose, CA, USA, March 2007. DOI: 10.1109/CGO.2007.10 44

Raymond J. Hookway and Mark A. Herdeg. Digital FX!32: Combining emulation and binary translation. *Digital Technical Journal*, pages 3–12, February 1997. 38

Galen Hunt and Doug Brubacher. Detours: Binary interception of win32 functions. In *Proceedings of the 3rd USENIX Windows NT Symposium*, pages 135–143, Seattle, WA, USA, July 1999. 39

Aamer Jaleel, Robert S. Cohn, Chi-Keung Luk, and Bruce Jacob. Cmp$im: A Pin-based on-the-fly single/multi-core cache simulator. In *Proceedings of the 2008 Workshop on Modeling, Benchmarking and Simulation*, MOBS '08, Beijing, China, June 2008. 34

Rahul Joshi, Michael D. Bond, and Craig Zilles. Targeted path profiling: Lower overhead path profiling for staged dynamic optimization systems. In *Proceedings of the 2nd Annual IEEE/ACM International Symposium on Code Generation and Optimization*, CGO '04, pages 239–250, Palo Alto, CA, USA, March 2004. 43

Minjang J. Kim, Chi-Keung Luk, and Hyesoon Kim. Prospector: Discovering parallelism via dynamic data-dependence profiling. Technical Report TR-2009-001, Georgia Institute of Technology, 2009. 23

Vladimir Kiriansky, Derek Bruening, and Saman Amarasinghe. Secure execution via program shepherding. In *Proceedings of the 11th USENIX Security Symposium*, pages 191–206, San Francisco, CA, USA, August 2002. 29

James R. Larus and Eric Schnarr. Eel: machine-independent executable editing. In *Proceedings of the ACM SIGPLAN Conference on Programming Language Design and Implementation*, PLDI '95, pages 291–300, La Jolla, CA, USA, 1995. ACM. DOI: 10.1145/223428.207163 55

Chi-Keung Luk, Robert Cohn, Robert Muth, Harish Patil, Artur Klauser, Geoff Lowney, Steven Wallace, Vijay Janapa Reddi, and Kim Hazelwood. Pin: Building customized program analysis tools with dynamic instrumentation. In *Proceedings of the ACM SIGPLAN Conference on Programming Language Design and Implementation*, PLDI '05, pages 190–200, Chicago, IL, USA, June 2005. DOI: 10.1145/1065010.1065034 58

Jonas Maebe, Michiel Ronsse, and Koen De Bosschere. Diota: Dynamic instrumentation, optimization, and transformation of applications. In *Proceedings of the 4th Workshop on Binary Translation*, WBT '02, Charlottesville, VA, USA, September 2002. 55

Peter S. Magnusson, Magnus Christensson, Jesper Eskilson, Daniel Forsgren, Gustav Hållberg, Johan Högberg, Fredrik Larsson, Andreas Moestedt, and Bengt Werner. Simics: A full system simulation platform. *IEEE Computer*, 35(2):50–58, February 2002. DOI: 10.1109/2.982916 55

Duane Merrill and Kim Hazelwood. Trace fragment selection within method-based JVMs. In *Proceedings of the 4th Annual ACM SIGPLAN/SIGOPS International Conference on Virtual Execution Environments*, VEE '08, pages 41–50, Seattle, WA, USA, March 2008. DOI: 10.1145/1346256.1346263 44

Tipp Moseley, Alex Shye, Vijay Janapa Reddi, Dirk Grunwald, and Ramesh V. Peri. Shadow profiling: Hiding instrumentation costs with parallelism. In *Proceedings of the 5th Annual IEEE/ACM International Symposium on Code Generation and Optimization*, CGO '07, San Jose, CA, USA, March 2007. DOI: 10.1109/CGO.2007.35 53

Satish Narayanasamy, Cristiano Pereira, Harish Patil, Robert Cohn, and Brad Calder. Automatic logging of operating system effects to guide application-level architecture simulation. In *Proceedings of the Joint International Conference on Measurement and Modeling of Computer Systems*, SIGMETRICS '06/Performance '06, pages 216–227, Saint Malo, France, June 2006. DOI: 10.1145/1140103.1140303 24

Nicholas Nethercote. *Dynamic Binary Analysis and Instrumentation*. PhD thesis, University of Cambridge, Cambridge, U.K., November 2004. 58

Nicholas Nethercote and Julian Seward. Valgrind: a framework for heavyweight dynamic binary instrumentation. In *Proceedings of the ACM SIGPLAN Conference on Programming Language Design and Implementation*, PLDI '07, pages 89–100, San Diego, CA, USA, June 2007. DOI: 10.1145/1273442.1250746 58

Heidi Pan, Krste Asanovic, Robert Cohn, and Chi-Keung Luk. Controlling program execution through binary instrumentation. In *Proceedings of the Workshop on Binary Instrumentation and Applications*, WBIA '05, St. Louis, MO, USA, September 2005. DOI: 10.1145/1127577.1127587 48

Harish Patil, Robert Cohn, Mark Charney, Rajiv Kapoor, Andrew Sun, and Anand Karunanidhi. Pinpointing representative portions of large intel itanium programs with dynamic instrumentation. In *Proceedings of the 37th Annual IEEE/ACM International Symposium on Microarchitecture*, MICRO-37, pages 81–92, Portland, OR, USA, December 2004. DOI: 10.1109/MICRO.2004.28 32

Harish Patil, Cristiano Pereira, Mack Stallcup, Gregory Lueck, and James Cownie. PinPlay: A framework for deterministic replay and reproducible analysis of parallel programs. In *Proceedings of the 8th Annual IEEE/ACM International Symposium on Code Generation and Optimization*, CGO '10, pages 2–11, Toronto, Ontario, Canada, April 2010. DOI: 10.1145/1772954.1772958 24

Vijay Janapa Reddi, Dan Connors, Robert Cohn, and Michael D. Smith. Persistent code caching: Exploiting code reuse across executions and applications. In *Proceedings of the 5th Annual International IEEE/ACM Symposium on Code Generation and Optimization*, CGO '07, pages 74–88, San Jose, CA, USA, March 2007. DOI: 10.1109/CGO.2007.29 8

Ted Romer, Geoff Voelker, Dennis Lee, Alec Wolman, Wayne Wong, Hank Levy, Brian Bershad, and Brad Chen. Instrumentation and optimization of win32/intel executables using etch. In *Proceedings of the USENIX Windows NT Workshop*, Seattle, WA, USA, 1997. USENIX Association. 55

Kevin Scott, Naveen Kumar, Siva Velusamy, Bruce Childers, Jack W. Davidson, and Mary Lou Soffa. Retargetable and reconfigurable software dynamic translation. In *Proceedings of the 1st Annual IEEE/ACM International Symposium on Code Generation and Optimization*, CGO '03, pages 36–47, San Francisco, CA, USA, March 2003. DOI: 10.1109/CGO.2003.1191531 45, 55

Alex Skaletsky, Tevi Devor, Nadav Chachmon, Robert S. Cohn, Kim Hazelwood, Vladimir Vladimirov, and Moshe Bach. Dynamic program analysis of microsoft windows applications. In *Proceedings of the IEEE International Symposium on Performance Analysis of Systems and Software*, ISPASS '10, pages 2–12, White Plains, NY, USA, March 2010. DOI: 10.1109/ISPASS.2010.5452079 41, 52

James E. Smith and Ravi Nair. *Virtual Machines: Versatile Platforms for Systems and Processes*. Morgan Kaufmann, June 2005. 2, 58

Swaroop Sridhar, Jonathan S. Shapiro, Eric Northup, and Prashanth P. Bungale. HDTrans: An open source, low-level dynamic instrumentation system. In *Proceedings of the 2nd Annual ACM SIGPLAN/SIGOPS International Conference on Virtual Execution Environments*, VEE '06, pages 175–185, Ottawa, Ontario, Canada, June 2006. DOI: 10.1145/1134760.1220166 55

Amitabh Srivastava and Alan Eustace. Atom: a system for building customized program analysis tools. In *Proceedings of the ACM SIGPLAN Conference on Programming Language Design and Implementation*, PLDI '94, pages 196–205, Orlando, FL, USA, June 1994. ACM. DOI: 10.1145/989393.989446 55

Gang-Ryung Uh, Robert Cohn, Bharadwaj Yadavalli, Ramesh Peri, and Ravi Ayyagari. Analyzing dynamic binary instrumentation overhead. In *Proceedings of the Workshop on Binary Instrumentation and Applications*, WBIA '06, San Jose, CA, USA, October 2006. 4

Steven Wallace and Kim Hazelwood. SuperPin: Parallelizing dynamic instrumentation for real-time performance. In *Proceedings of the 5th Annual IEEE/ACM International Symposium on Code Generation and Optimization*, CGO '07, pages 209–217, San Jose, CA, USA, March 2007. DOI: 10.1109/CGO.2007.37 52

Emmett Witchel and Mendel Rosenblum. Embra: Fast and flexible machine simulation. In *Proceedings of the ACM SIGMETRICS International Conference on Measurement and Modeling of Computer Systems*, SIGMETRICS '96, pages 68–79, Philadelphia, PA, USA, 1996. ACM. DOI: 10.1145/233008.233025 55

Xiaolan Zhang, Zheng Wang, Nicholas Gloy, J. Bradley Chen, and Michael D. Smith. System support for automatic profiling and optimization. In *Proceedings of the Sixteenth ACM Symposium on Operating Systems Principles*, SOSP '97, pages 15–26, Saint Malo, France, 1997. ACM. DOI: 10.1145/268998.266640 55

Qin Zhao, Ioana Cutcutache, and Weng-Fai Wong. Pipa: Pipelined profiling and analysis on multicore systems. *ACM Transactions on Architecture and Code Optimization*, 7(3):13:1–13:29, December 2010. DOI: 10.1145/1880037.1880038 53

Author's Biography

KIM HAZELWOOD

Kim Hazelwood is an Assistant Professor of Computer Science at the University of Virginia and a faculty consultant for Intel Corporation. She works at the boundary between hardware and software, with research efforts focusing on computer architecture, run-time optimizations, and the implementation and applications of process virtualization systems. She received the Ph.D. degree from Harvard University in 2004. Since then, she has become widely known for her active contributions to the Pin dynamic instrumentation system, which allows users to easily inject arbitrary code into existing program binaries at run time (`www.pintool.org`). Pin is widely used throughout industry and academia to investigate new approaches to program introspection, optimization, security, and architectural design. It has been downloaded over 50,000 times and cited in over 800 publications since it was released in July 2004. Kim has published over 40 peer-reviewed articles related to computer architecture and virtualization. She has served on over two dozen program committees, including ISCA, PLDI, MICRO, OSDI, and PACT, and was a program chair of CGO 2010. Kim is the recipient of numerous awards, including the FEST Distinguished Young Investigator Award for Excellence in Science and Technology, an NSF CAREER Award, a Woodrow Wilson Career Enhancement Fellowship, the Anita Borg Early Career Award, an MIT Technology Review "Top 35 Innovators under 35 Award", and research awards from Microsoft, Google, NSF, and the SRC. Her research has been featured in MIT Technology Review, Computer World, ZDNet, EE Times, and Slashdot.

Printed in the United States
by Baker & Taylor Publisher Services